Disturbance Rejection Control

Edited by Mohammad Shamsuzzoha and G. Lloyds Raja

Published in London, United Kingdom

Disturbance Rejection Control
http://dx.doi.org/10.5772/intechopen.105256
Edited by Mohammad Shamsuzzoha and G. Lloyds Raja

Contributors
Xinye You, Rodrigo de Figueiredo, Bruno Toso, Jean Schmith, Jenq-Lang Wu, Arumugam Arunkumar, Samira Chebli, Mohammad Shamsuzzoha, G. Lloyds Raja

First published in London, United Kingdom, 2023 by IntechOpen
IntechOpen is the global imprint of INTECHOPEN LIMITED, registered in England and Wales, registration number: 11086078, 5 Princes Gate Court, London, SW7 2QJ, United Kingdom

British Library Cataloguing-in-Publication Data
A catalogue record for this book is available from the British Library

Additional hard and PDF copies can be obtained from orders@intechopen.com

Disturbance Rejection Control
Edited by Mohammad Shamsuzzoha and G. Lloyds Raja
p. cm.
Print ISBN 978-1-83768-801-2
Online ISBN 978-1-83768-802-9
eBook (PDF) ISBN 978-1-83768-803-6

For EU product safety concerns:
IN TECH d.o.o., Prolaz Marije Krucifikse Kozulić 3, 51000 Rijeka, Croatia,
info@intechopen.com or visit our website at intechopen.com.

Meet the editors

Dr. M. Shamsuzzoha has a Ph.D. in Chemical Engineering with a specialization in process design, modeling, simulation, and control. He conducts studies and provides process engineering, modeling, and simulation support to clients in a wide range of process industries. Prior to his present position, Dr. Shamsuzzoha worked in ADNOC Refining Co. as a Senior Specialist in Process Engineering focused on technical support to refinery and gas plants, including process modeling and simulation as well as modification and design of refinery units and associated gas plants. Recently, he completed an MBA in Entrepreneurship and Leadership. Dr. Shamsuzzoha has unique work experience in academics, industries, and research and development in a multicultural environment. He is a pioneering researcher and practicing engineering in process modelling, simulation, and control.

Dr. G. Lloyds Raja received his bachelor's and master's degrees in Engineering from Anna University, India. He was awarded a Ph.D. from the Indian Institute of Technology Patna for his dissertation entitled "Enhanced Cascade Control Strategies for Unstable and Industrial Processes". For a brief period, Dr. Raja was a postdoctoral researcher at the Department of Automation, Shanghai Jiao Tong University, China. Presently, he is an assistant professor at the National Institute of Technology Patna, India. He has more than 10 years of teaching and research experience. He has authored/co-authored nearly forty papers in international journals and peer-reviewed conference proceedings. He has taught courses on linear control systems, modern control theory, electronic circuits, and industrial process control and delivered invited lectures in several forums.

Contents

Preface

Meeting the continuously evolving challenges in industrial process control requires efficient control strategies that can yield improved performance compared to contemporary methods. In industrial process control, disturbance rejection is more vital than setpoint tracking. This is because in scenarios like manufacturing processes, the production rate is hardly altered, whereas the plant is prone to encounter disturbance at any instance. This book covers a range of topics within the realm of disturbance rejection control for various industrial processes.

Chapter 1 introduces the theme of the book, discussing the different types of disturbance rejection control techniques and their applications at a surface level.

Auto-tuning PID controllers have attracted much attention in recent times among control engineers. Though there are several analytical techniques existing for auto-tuning PID controllers, Chapter 2 presents a genetic algorithm-based heuristic approach. This auto-tuning procedure is generic and extendable to various classes of industrial processes. Controllers have also found applications in enhancing network efficiency.

Chapter 3 presents an ant colony optimizer-tuned PI controller for the aforementioned purpose. The findings in this chapter are validated through extensive simulations.

Chapter 4 presents an observer-based disturbance rejection technique for switched nonlinear networked systems under an event-triggered scheme. The chapter provides an observer-based, event-triggered disturbance rejection controller that ensures the resulting closed-loop form of the examined systems is exponentially stable.

Chapter 5 presents a robust control technique for space robots by considering friction characteristics. It gives an overview of space robots, including modeling the robot and designing a controller based on the obtained model.

My sincere thanks to all the chapter authors and Author Service Manager Ana Cink at IntechOpen who provided excellent support throughout the publication process. I believe that the contents of this book will help engineers and researchers working in the domain of PID controller design for various applications. We also acknowledge our respective organizations (Billington Process Technology and National Institute of Technology Patna) for providing us with the necessary facilities and infrastructure during this editorial task. Also, it would have been impossible to successfully edit this book without the support and cooperation of our families.

Dr. Mohammad Shamsuzzoha
Principal Process Engineer,
Billington Process Technology,
Sandvika, Norway

Dr. G. Lloyds Raja
Assistant Professor,
Department of Electrical Engineering,
National Institute of Technology Patna,
Patna, India

Chapter 1

Introductory Chapter: Introduction to Disturbance Rejection Control

G. Lloyds Raja and Mohammad Shamsuzzoha

1. Introduction

Disturbing dynamic systems on a regular basis can have a significant impact on their performance and stability. Disturbance rejection control techniques seek to mitigate the impact of disturbances while preserving desirable system behavior. This chapter delves further into the definition, goals, control mechanisms, and applications of disturbance rejection control. The theoretical foundations and practical applications of various disturbance rejection control systems are also discussed, with an emphasis on the importance of robustness and adaptability. Any dynamic system will experience disturbances, which can be caused by both internal and external uncertainties. These disruptions have a substantial impact on the system's performance and can cause it to deviate from target setpoints or trajectories [1]. Disturbance rejection control strategies are used to lessen the effects of disturbances and preserve desirable system behavior. An overview of disturbance rejection control and its significance in many applications will be given in this introduction. This chapter concludes by summarizing the prospects for the future of this field and prospective future research endeavors.

2. Objectives and definition

Disturbance rejection control is a subfield of control engineering that aims to reduce the impact of disturbances on the output or performance of the system [2]. In the face of disturbances, the major goal of disturbance rejection control is to keep the system stable, accurate, and resilient. The control system strives to make sure that the output closely resembles the desired reference signal or trajectory, even in the presence of disturbances, by actively compensating for them [3].

3. Types of disturbances

Depending on the characteristics of the system and its surroundings, disturbances can take many distinct forms. Internal and external disturbances are two major groups into which they might be divided. External disturbances come from the environment and can be caused by things like temperature fluctuations, wind loads, or adjustments to the input signals [4]. On the other side, internal disturbances result from uncertainties within the system itself, such as parameter changes, sensor noise, or model errors [5, 6].

IntechOpen

4. Techniques for disturbance rejection control

To reject disturbances and preserve desirable system performance, a variety of control strategies are used. Among the methods that are frequently utilized are:

4.1 Feedforward control

This technique uses an estimated disturbance model and a compensating control action to mitigate the effects of the disturbance before they have an impact on the system output. This method works best when the disturbance can be precisely quantified or predicted in advance [7–11].

4.2 Feedback control

The classic strategy of feedback control compares the system's output continually to the intended reference signal and modifies the control action to minimize error. Feedback control aids in making up for disturbances that cannot be precisely predicted or measured in advance [12–28].

4.3 Adaptive control

Adaptive control approaches modify the control action in response to shifting system dynamics and disturbance characteristics by using online parameter estimation and adaptation algorithms. In situations when the system parameters or disturbance characteristics change over time, adaptive control is especially helpful [29–34].

5. Robustness and adaptability

Techniques for disturbance rejection control must be resilient and adaptable to provide efficient disturbance compensation. The term "robustness" describes a system's capacity to continue operating consistently and accurately in the face of unknowns and interruptions. Robust design methodologies and uncertainty modeling are used into robust control algorithms to ensure stability and performance guarantees even when disturbances are greater than expected [12].

A control system's adaptability is its capacity to change its structure or parameters in response to varying environmental factors. Using adaptive control techniques, the system may continually estimate and update the properties of the disturbance and modify the control action as necessary. In the face of time-varying disturbances or shifting system dynamics, this adaptability aids in maintaining system performance [7].

6. Applications

Disturbance rejection control is used in a variety of industries, including process, manufacturing, robotics, and aerospace.

6.1 Aerospace

Disturbance rejection management is essential in aerospace applications for stabilizing aircraft during turbulence and fending off outside disturbances like wind gusts. By successfully adjusting for disturbances that impact the aircraft's trajectory and performance, it provides safe and stable flight [35].

6.2 Robotics

In order to maintain exact positioning and tracking of robotic arms in the face of external forces, uncertainties, or disturbances, disturbance rejection control is crucial. It makes it possible for robots to complete jobs reliably and precisely even in changing environments [35].

6.3 Manufacturing

To obtain accurate control of diverse operations, disturbance rejection control is used in manufacturing processes. For instance, disturbance rejection control in CNC machining helps maintain precise tool positioning and tracking by adjusting for outside influences or uncertainties that can impair the quality of the cutting [36].

6.4 Process industries

Maintaining product quality and stability is crucial in process industries, such as chemical plants, therefore, disturbance rejection control is essential. In order to maintain constant process performance and product quality, it enables the control system to account for disturbances, variations in input parameters, or uncertainties [17].

7. Motivation for this book

Disturbance rejection control is a fundamental aspect of control engineering that addresses the challenges posed by disturbances in dynamic systems. By employing various control techniques such as feedforward, feedback, and adaptive control, disturbance rejection control mitigates the effects of disturbances and ensures stable system behavior. The robustness and adaptability of control algorithms are essential to handle uncertainties and changing conditions effectively. Disturbance rejection control has broad applications in aerospace, robotics, manufacturing, and process industries, enabling systems to operate reliably and achieve desired performance even in the presence of disturbances. Future research in this field should focus on developing advanced disturbance modeling techniques, robust control algorithms, and adaptive strategies to enhance disturbance rejection capabilities and address emerging challenges.

Author details

G. Lloyds Raja[1] and Mohammad Shamsuzzoha[2*]

1 Department of Electrical Engineering, National Institute of Technology Patna, India

2 Billington Process Technology, Sandvika, Norway

*Address all correspondence to: smzoha@gmail.com

References

[1] Astrom KJ, Wittenmark B. Adaptive Control. New York: Dover Publications; 2013

[2] Chen W, Francis B. Optimal Robust Control: Linear Matrix Inequalities in Control. The Netherlands: Dover Publications; 2005

[3] Miyamoto K, Nakano S, She J, Sato D, Chen Y, Han QL. Design method of tuned mass damper by linear-matrix-inequality-based robust control theory for seismic excitation. Journal of Vibration and Acoustics. 2022;**144**(4):041008

[4] Goodwin GC, Graebe SF, Salgado ME. Control System Design. Australia: Prentice Hall; 2013

[5] Dorato P. Disturbance rejection in nonlinear control systems. Automatica. 1981;**17**(5):693-699

[6] Wood AJ, Wollenberg BF, Sheblé GB. Power Generation, Operation, and Control. USA: John Wiley & Sons; 2013

[7] Shamsuzzoha M, Raja GL. Introductory chapter: PID-based industrial process control. In: PID Control for Linear and Nonlinear Industrial Processes. London, UK: IntechOpen; 2023

[8] Mehta U, Aryan P, Raja GL. Tri-parametric fractional-order controller Design for Integrating Systems with time delay. IEEE Transactions on Circuits and Systems II: Express Briefs. 2023. DOI: 10.1109/TCSII.2023.3269819 [In press]

[9] Kumar D, Raja GL, Arrieta O, Vilanova R. Fractional-order model identification and indirect internal model controller design for higher-order processes. In: Proceedings of IFAC World Congress. Yokohama, Japan: Japan IFAC-PapersOnLine; 2023

[10] Shamsuzzoha M, Lee M. IMC-PID controller design for improved disturbance rejection of time-delayed processes. Industrial & Engineering Chemistry Research. 2007;**46**(7):2077-2091

[11] Dudhe S, Raja GL, Dheer DK. Modeling and control of suction pressure in portable meconium aspirator system using fractional order IMC PID controller and RDR techniques. In: Materials Today: Proceedings. Meerut, India: Elsevier; 2023. DOI: 10.1016/j.matpr.2023.01.404 [In press]

[12] Raja GL. Robust I-PD controller design with case studies on boiler steam drum and bioreactor. In: 2023 15th International Conference on Computer and Automation Engineering (ICCAE). Sydney, Australia: IEEE; 2023. pp. 486-491

[13] Aryan P, Raja GL, Vilanova R, Meneses M. Repositioned internal model control strategy on time-delayed industrial processes with inverse behavior using equilibrium optimizer. IEEE Access. 2023;**11**:54556-54568. DOI: 10.1109/ACCESS.2023.3281691

[14] Aryan P, Raja GL, Vilanova R. Experimentally verified optimal bi-loop re-located IMC strategy for unstable and integrating systems with dead time. International Journal of Systems Science. 2023;**54**(7):1531-1549. DOI: 10.1080/00207721.2023.2180782

[15] Das D, Chakraborty S, Raja GL. Enhanced dual-DOF PI-PD control of integrating-type chemical processes.

International Journal of Chemical Reactor Engineering. 2022. DOI: 10.1515/ijcre-2022-0156 [In press]

[16] Mukherjee D, Raja GL, Kundu P, Ghosh A. Improved fractional augmented control strategies for continuously stirred tank reactors. Asian Journal of Control. 2023;**25**(3): 2165-2182

[17] Raja GL, Ali A. New PI-PD controller design strategy for industrial unstable and integrating processes with dead time and inverse response. Journal of Control, Automation and Electrical Systems. 2021;**32**(2):266-280

[18] Kumari S, Aryan P, Raja GL. Design and simulation of a novel FOIMC-PD/P double-loop control structure for CSTRs and bioreactors. International Journal of Chemical Reactor Engineering. 2021;**19**(12):1287-1303

[19] Kumar D, Aryan P, Raja GL. Design of a novel fractional-order internal model controller-based smith predictor for integrating processes with large dead-time. Asia-Pacific Journal of Chemical Engineering. 2022;**17**(1):e2724

[20] Kumar D, Raja GL. Unified fractional indirect IMC-based hybrid dual-loop strategy for unstable and integrating type CSTRs. International Journal of Chemical Reactor Engineering. 2022;**21**(3):251-272

[21] Kumar D, Aryan P, Raja GL. Decoupled double-loop FOIMC-PD control architecture for double integral with dead time processes. The Canadian Journal of Chemical Engineering. 2022;**100**(12):3691-3703

[22] Aryan P, Raja GL. A novel equilibrium optimized double-loop control scheme for unstable and integrating chemical processes involving dead time. International Journal of Chemical Reactor Engineering. 2022;**1**:20

[23] Kumari S, Aryan P, Kumar D, Raja GL. Hybrid dual-loop control method for dead-time second-order unstable inverse response plants with a case study on CSTR. International Journal of Chemical Reactor Engineering. 2022;**1**:11

[24] Raja GL, Ali A. Series cascade control: An outline survey. In: 2017 Indian Control Conference (ICC). Guwahati, India: IEEE; 2017. pp. 409-414

[25] Raja GL, Ali A. Modified parallel cascade control strategy for stable, unstable and integrating processes. ISA Transactions. 2016;**65**:394-406

[26] Raja GL, Ali A. Enhanced tuning of smith predictor based series cascaded control structure for integrating processes. ISA Transactions. 2021;**114**:191-205

[27] Raja GL, Ali A. Smith predictor based parallel cascade control strategy for unstable and integrating processes with large time delay. Journal of Process Control. 2017;**52**:57-65

[28] Aryan P, Raja G. Equilibrium-optimized IMC-PD double-loop control strategy for industrial processes with dead time. In: Recent Advances in Mechanical Engineering. Singapore: Springer; 2023. pp. 37-50 Engineering Science, 232, 116322

[29] Anandh P, Aryan NK, Raja GL. Type-2 fuzzy-based branched controller tuned using arithmetic optimizer for load frequency control. Energy Sources, Part A: Recovery, Utilization, and Environmental Effects. 2022;**44**(2):4575-4596. DOI: 10.1080/15567036.2022.2078444

[30] Aryan P, Raja GL. Analysis of type-2 fuzzy I λ D μ-P controller for LFC with communication delay. In: 2022 IEEE Global Conference on Computing, Power and Communication Technologies (GlobConPT). IEEE; Sep 2022. pp. 1-7

[31] Ahmad S, Ali A. On active disturbance rejection control in presence of measurement noise. IEEE Transactions on Industrial Electronics. 2021;**69**(11):11600-11610

[32] Anand A, Kumari N, Aryan P, Raja GL. EO optimized novel Type-2 fuzzy ID-P controller for LFC of deregulated multi-area power system with robust stability analysis. In: IEEE Second International Conference on Power, Control and Computing Technologies ICPC^{2}T. Chhattisgarh: NIT Raipur; 2022. DOI: 10.1109/ICPC2T53885.2022.9776841

[33] Aryan P, Raja GL. Restructured LFC scheme with renewables and EV penetration using novel QOEA optimized parallel fuzzy I-PID controller. IFAC-PapersOnLine. 2022;**55**(1):460-466. DOI: 10.1016/j.ifacol.2022.04.076

[34] Aryan P, Raja GL. Design and analysis of novel QOEO optimized parallel fuzzy FOPI-PIDN controller for restructured AGC with HVDC and PEV. IJST Transactions of Electrical Engineering. 2022;**46**(2):565-587. DOI: 10.1007/s40998-022-00484-7

[35] Niku SB. Introduction to Robotics: Analysis, Control, Applications. USA: John Wiley & Sons; 2020

[36] Vilanova R, Visioli A. PID Control in the Third Millennium. London: Springer; 2012

Chapter 2

Auto-Tuning PID Controller Based on Genetic Algorithm

Rodrigo de Figueiredo, Bruno Toso and Jean Schmith

Abstract

The PID controller is widely used in industry and its tuning is always a concern for the plant stabilization. Several methods for auto-tuning the PID have been proposed over the years, however, the relay method is the most used even though this method may determine nonideal PID gains and cause some physical stress on the plant. Here is presented a proposal for an auto-tuning PID controller based on a genetic algorithm. Genetic algorithm is a well-known method that imitates the natural selection process in order to obtain approximate solutions to optimization problems. Here, the method is presented in underdamped plants with the hypothesis that any plant can be approximated to a second-order function. From the unit step response of the system, the maximum overshoot and peak time were used in the GA evolution to obtain optimal PID parameters. The system was tested with a set of parameters and compared to MATLAB PID tuner function. Using the rising time and the settling time of unit step response from the closed loop system as validation parameters, the GA presented better results than the MATLAB tuner for most cases.

Keywords: auto-tuning PID, genetic algorithm, artificial intelligence, PID controller, underdamped second order plants

1. Introduction

The PID (Proportional–Integral–Derivative) controller is widely used in industry for process control and robotics. The auto-tuning is an important feature embedded in some commercial PID controllers. This feature is responsible for tuning the PID controller by using the plant's unit step response and changing the *Kp*, *Ki* e *Kd* gains in order to obtain the plant stability [1–3]. The performance characteristics of a control system are often specified in terms of the transient response to unit step input. The transient response of a stable physical system frequently presents damped oscillations before achieving its steady state. The specification of the transient-response characteristics of a control system to unit step input is given by the determination of the *Td* delay time, *Tr* rise time, *Tp* peak time, *Mp* maximum overshoot, and *Ts* settling time.

Figure 1 presents a typical unit step response from a second-order system with the parameters indication. The *Td* represents the time from zero until the amplitude of the response reaches 0.5, the Tr is the time to the response to reach the amplitude of 1.0, and the *Tp* is the time to the response to reach the maximum overshoot *Mp*.

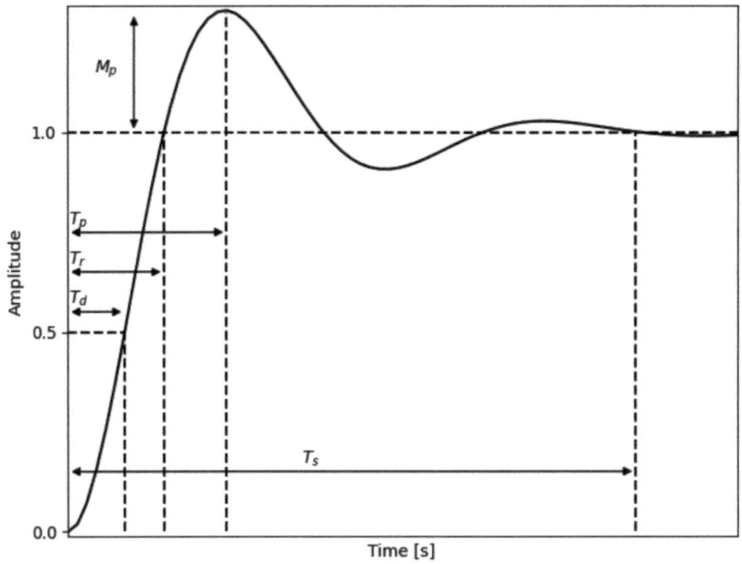

Figure 1.
Unit step response of a second-order plant with the parameters Td, Tr, Tp, Ts, and Mp indication.

At last, but not the least, the *Ts* is the time to the response to achieve the steady state condition. It is noteworthy that *Ts* depends on the chosen allowable oscillation tolerance in a steady-state condition. Although the parameters were presented for an under-dumped example, they are applicable to critically damped and over damped, as well. Therefore, the same procedure may be applied to any second-order plant [4, 5].

The PID controllers are tuned by the *Kp* proportional, *Ki* integral, and *Kd* derivative gains and are directly responsible for the system stability. A well-tuned PID controller works to decrease the oscillation in the system. The stability of the system may impact directly the productivity of a manufacturing process or the precision of a robot, for example. Thus, the tuning and auto-tuning methods are important, as evidenced by works [2, 6].

There are well-established auto-tuning algorithms that implement the relay and Ziegler-Nichols methods [7]. The Ziegler-Nichols method requires the parameters from the system in an oscillatory fashion to obtain the critic *Ku* gain and the oscillation period *Tu* to compute the PID gains. However, this procedure may cause some stress on the plant. To avoid that issue, the relay method is an alternative to the parameter extraction procedure. This method is based on restricting the controller's output in order to create controlled oscillation on the plant. The procedure yields the same two parameters obtained from the previously mentioned method and used to obtain the PID gains. It consists of replacing the controller with a relay, which operates with a hysteresis and limited output. Although the relay method may be less stressful to the system, oscillatory behavior is still needed for a certain amount of time, which may be impractical for certain plants [8].

Another approach is the usage of artificial intelligence [9]. Applications using Fuzzy Logic [10] for PID tuning are based on a predetermined expert system, which uses a rule table for each of the controller parameters. By using this set of rules, it was possible to reach a satisfactory result. Other works using fuzzy [11–13] and flower pollination algorithms [14] presented good results for obtaining the PID parameters. The point here is that auto-tuning of PID controllers using artificial intelligence is a possibility.

The artificial neural network (ANN) has the disadvantage in the need of a large quantity of data from the plant. Whereas, the use of genetic algorithm (GA) uses only the unit step response parameters. The GA is able to produce better results than Ziegler-Nichols in terms of performance [15, 16]. Therefore, this work proposes the usage of GA for tuning the values of Kp, Ki, and Kd gains from the unit step parameters Mp and Ts. The novelty here is the usage of a single-unit step response to the closed system for the PID tuning.

This method allows the tuning of the PID controller with minimal physical plant stress. The method allows the parameter computation from any physical plant of unknown mathematical model since the idea is the approximation to a second-order function.

2. Methodology

The proposed system uses the extraction of temporal parameters from the plant unit step response. The method uses the extracted temporal data to apply it to a transfer function. This transfer function (second-order mathematical model) was used by the GA to validate the parameters Kp, Ki, and Kd and its result is compared with an ideal one, which is a function with a stable response in the set point, low overshoot, and low ripple. **Figure 2** presents the block diagram for the proposed system.

A unit step signal is applied to the plant. Next, from the unit step response of the plant is possible to obtain the peak time Tp and maximum overshoot . For the GA design, the maximum ripple tolerance in steady state condition assumed for Ts determination was 0.5%. This 0.5% ripple is, on purpose, a restricted value in relation to the values assumed in control plants for regulatory response [17–19]. Further, the damping factor ξ and the undamped natural frequency ω_n are computed by Eqs. (1) and (2), respectively.

$$\xi = \sqrt{\frac{\left(\frac{\log(M_P)}{\pi}\right)^2}{1+\left(\frac{\log(M_P)}{\pi}\right)^2}} \tag{1}$$

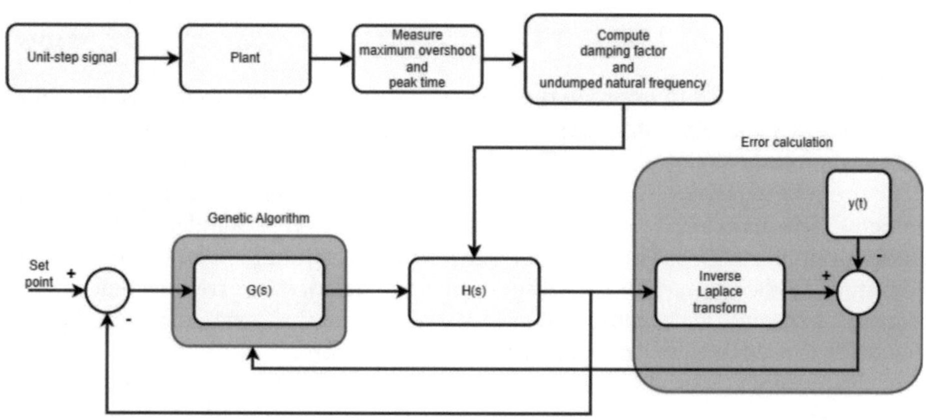

Figure 2.
Block diagram of the proposed method.

$$\omega_n = \frac{\pi}{\left(T_P.\sqrt{1-\xi^2}\right)} \tag{2}$$

Hence, the idea is to approximate any plant to a second-order system. By the ω_n and ξ computation it is possible to determine the Laplace transformation of an approximate second-order transfer function as presented in Eq. (3), where s denotes the Laplace variable.

$$H_{(s)} = \frac{\omega_n^2}{s^2 + 2\xi\omega_n s + \omega_n^2} \tag{3}$$

The PID controller is determined by Eq. (4).

$$G_{(S)} = K_p + \frac{K_i}{s} + K_d s \tag{4}$$

The feedback model of the system yields Eq. (5).

$$\frac{Y_{(s)}}{X_{(s)}} = \frac{G_{(s)}H}{1 + G_{(s)}H} \tag{5}$$

At this point, the control structure is established. Now the question is how to determine the parameters to improve the $G(s)$ to produce a stable system. Again, the unit step signal is applied to the closed system. From the feedback model, the inverse Laplace transform was computed to obtain the unit step response of the controlled system. The error was computed by Eq. (6), where y is the desired output and a_0 is the output of the controlled system. The brackets denote the discrete form of the signal.

$$\xi = \sum_{i=0}^{n} \left(y[i] - a_0[i]\right)^2 \tag{6}$$

The ideal desired output $y(t)$ is computed by Eq. (7), in order to obtain a response approximated to a Heaviside function stable in the set point with low overshoot and low ripple.

$$y(t) = 1 - e^{100.T_P.t} \tag{7}$$

This error was used to improve the GA algorithm that computes new values for *Kp*, *Ki*, and *Kd* to achieve a stable output to the system. With this system evaluating the plant has minimal physical stress. The PID tuning was given by the GA evolution.

The genetic algorithm is a method to seek approximate solutions to optimization problems. The main idea is to look for better solutions from a population of abstract solution representations. The evolution starts with a random population to form the generation, in each generation, each solution is evaluated, some are selected, some are discarded, and some are recombined or mutated to form the next generation. The next generation is used in the next interaction of the algorithm.

In the presented method, the parameters to be tuned in the PID are the *Kp* proportional, *Ki* integrative, and *Kd* derivative gains. Therefore, the GA genome is created with the PID gains as its chromosomes. The chromosomes are the elements that evolve through the GA process. The GA genome is formed by the parameters *Kp*,

Ki, and *Kd*, as floating-point chromosomes. There was the possibility of establishing limits in variation of values of each chromosome of the genome, however, this practice limits the evolutionary capacity of the GA. Here it was decided not to create this limit and to work only on the hyperparameters of the GA. In this way, there is an evolutionary freedom of the parameters *Kp*, *Ki*, and *Kd*, not limiting the exploration.

The GA optimal hyperparameters were defined after testing. In these tests, the hyperparameters were varied individually over a well-known plant (a classic underdamped system of second order). We used some different plants to realize these tests and validate the system. Their step response was generated using MATLAB and fed into the system in order to analyze different overshoot levels. With these tests, the optimum hyperparameters were a population with 100 individuals (genomes), 50 generations (50 times that the GA has the opportunity to evolve), and a mutation chance of 6% (a random factor responsible to avoid locals optimum).

3. Discussion and results

In order to check the effectiveness of the proposed method for PID tuning, we compared the tuning capability with the pidtune function from MATLAB, which is a well-known tool for mathematics and very robust for control analysis. Our method was developed in MATLAB, as well, for a fair comparison.

Table 1 presents the results for the MATLAB pintune and the proposed GA method for a second-order plant with different values of ξ and ω_n. The chosen values for ξ were 0.1, 0.3, 0.6, and 0.9. For each ξ, the frequencies of 2 and 10 rad/s were used as ω_n. A steady-state error for all numerical experiments was not observed. The performance of each system was compared in terms of rising time *Tr* and settling time *Ts*.

Observing the *Tr* and *Ts*, in the most cases the proposed GA method reached better results than the MATLAB function. Considering the rising time *Tr*, the only situation that the MATLAB function presented a better result than GA was with ξ = 0, 6 and ω_n = 10 rad/s. In the case of *Ts*, the MATLAB achieved better results only in three situations, with ξ = 0.3, 0.6, and 0.9, in cases with ω_n = 10 rad/s. One might note that the situations that the GA did not achieve better results than the MATLAB function were with ω_n = 10 rad/s. This behavior may be explained by the chosen $y(t)$ function and the higher oscillatory aspect of the plant unit step response, which yields a higher error and, therefore, an additional difficulty to the GA correction of the *Kp*, *Ki*, and *Kd* gains.

Figure 3 presents the unit step response of the second order plant and the closed loop with the MATLAB function and the GA method with ξ = 0.1 and ω_n = 2 and 10 rad/s. One might note that we tacitly changed the scale of the graphics in order to present the most interesting behavior on each one. Note the better performance of the GA method against the MATLAB function. For the MATLAB function, ξ = 0.1 and ω_n = 2 rad/s, it was observed a maximum overshoot of 2%, for the other closed-loop systems with MATLAB function computations there was no overshoot.

Figure 4 presents the results for ξ = 0.3 and ω_n = 2 and 10 rad/s. Note the worst results of GA for ω_n = 10 rad/s, as discussed before, taking a long period to achieve the stability. On the other hand, for ω_n = 2 rad/s, the GA method presented better results.

Figure 5 presented the results for ξ = 0.6 and ω_n = 2 and 10 rad/s. For ω_n = 2 rad/s, the GA method achieved better results and for ω_n = 10 rad/s the MATLAB function performed better. For ω_n = 10 rad/s the GA method presented an overshoot of 8%.

ξ	ω_n (rad/s)	MATLAB	GA	MATLAB	GA	MATLAB	GA	MATLAB	GA	MATLAB	GA
		K_P	K_P	K_I	K_I	K_D	K_D	tr (ms)	tr (ms)	ts (ms)	ts (ms)
0.1	2	24.20	24.445	11.100	98.865	13.100	39.782	41.0	**14.4**	61.0	**24.2**
	10	1.05	1.000	0.546	7.691	0.507	1.000	70.0	**25.0**	11200.0	**230.0**
0.3	2	24.20	59.656	11.100	99.494	13.100	39.942	43.0	**15.0**	68.0	**24.0**
	10	3.16	3.505	4.900	0.362	0.509	0.332	67.0	**64.0**	**1980.0**	29600.0
0.6	2	31.60	99.083	19.000	99.752	13.100	47.501	44.0	**12.0**	76.0	**21.0**
	10	6.31	1.000	19.400	13.833	0.514	1.000	**48.0**	571.0	**631.0**	2420.0
0.9	2	47.30	99.816	42.600	99.974	13.100	80.820	44.0	**7.0**	75.0	**13.0**
	10	9.47	9.858	42.900	99.310	0.525	1.000	45.0	**31.0**	**87.0**	215.0

Table 1.
Performance comparison between the results obtained by MATLAB pidtune and the proposed genetic algorithm. In bold are highlighted the better results for each parameter.

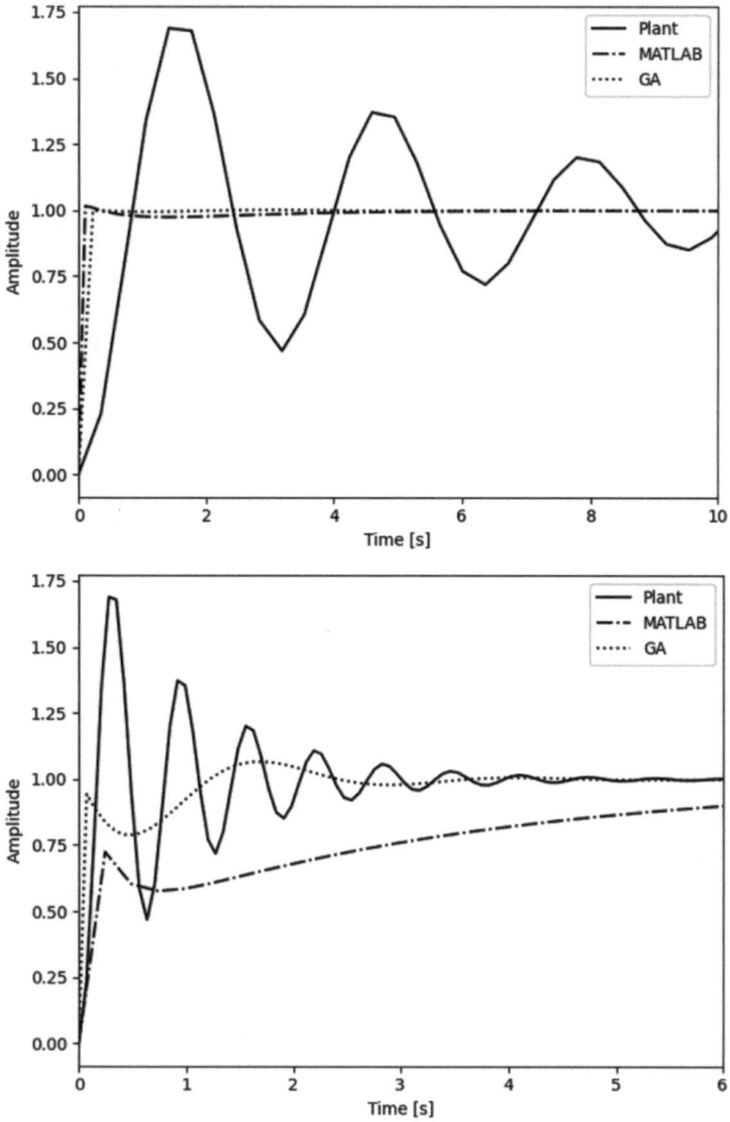

Figure 3.
Unit step response for ξ = 0.1 and (a) = 2 rad/s and (b) ω_n = 10 rad/s.

Finally, **Figure 6** presents the results for ξ = 0.9 and ω_n = 2 and 10 rad/s. Note, again, that for ω_n = 2 rad/s, the GA method presented better results than the MATLAB function, although, for ω_n = 10 rad/s, the MATLAB function presented better results. For ω_n = 10 rad/s, the GA presented an overshoot of 2%. For the other commented closed-loop systems with GA computation, there were no overshoots.

It was possible to observe that both techniques eliminated the overshoot and searched for the smallest rising time Tr, which are good features for a tuning system. As aforementioned, the GA method presented better results for the most cases and the cases that the MATLAB pidtune function presented better results were with higher oscillations of the plant unit step response, in which, with a better choice of $y(t)$ may

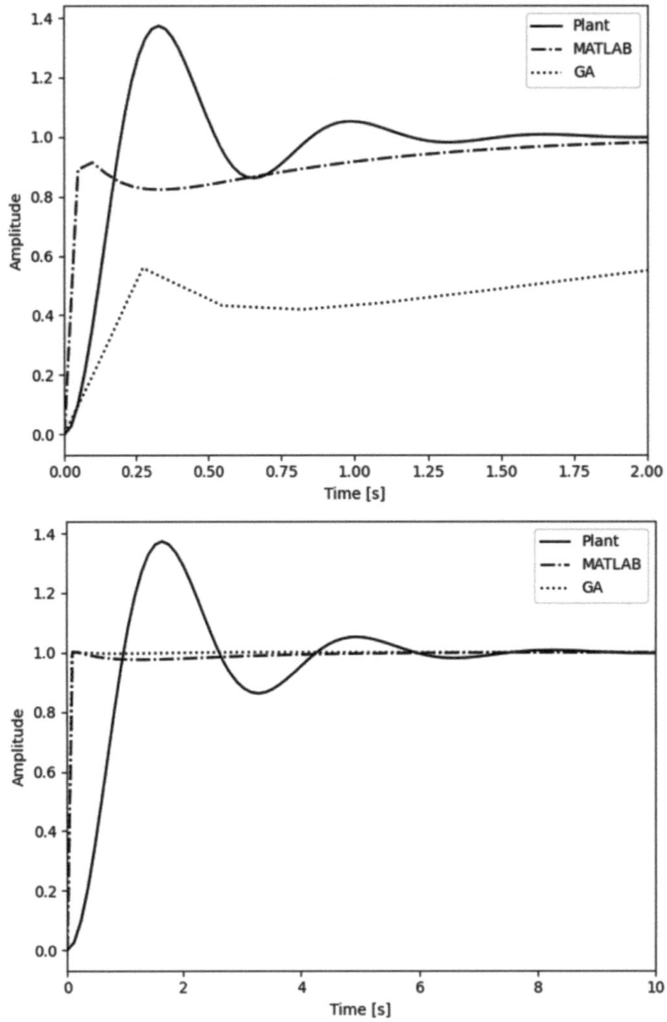

Figure 4.
Unit step response for $\xi = 0.3$ and (a) = 2 rad/s and (b) $\omega_n = 10$ rad/s.

mitigate this issue. The MATLAB function uses all available knowledge about control systems in order to provide optimal PID parameters, in addition to rising time Tr, settling time Ts, and overshoot Mp, it also uses stability and disturbance robustness. The GA proposed method uses only the error computation to compare the obtained response of the chromosome with an ideal unit step response. Therefore, the MATLAB function requires prior knowledge to configure the pidtune and the idea of the method proposed in this article is to depend as little as possible on the user's knowledge and avoid the stress of the plant.

Hence, the proposed method was developed to obtain the better performance in terms of Tr, Ts, and Mp. The disturbance robustness and other important features were not considered. The GA method presented good results in most cases, and it is possible to reach better results by adjusting the $y(t)$ function in order to consider more characteristics of the plant.

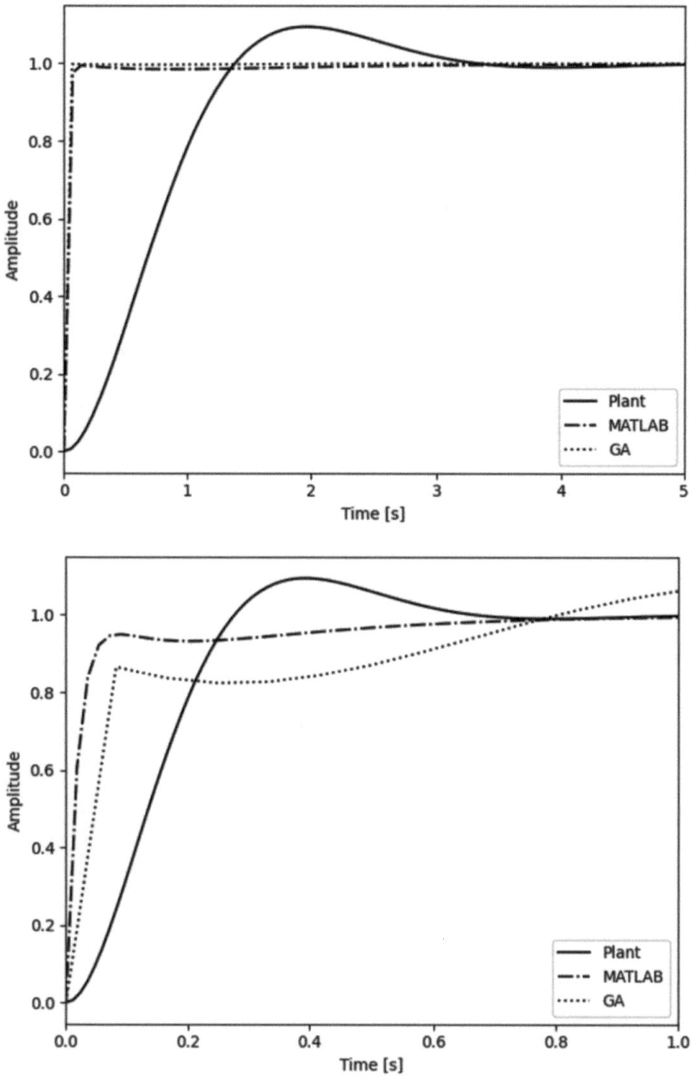

(a)

(b)

Figure 5.
Unit step response for ξ = 0.6 and (a) = 2 rad/s and (b) ω_n = 10 rad/s.

4. Conclusion

The presented work proposed a method for auto-tuning PID controllers using genetic algorithm. The method was applied to different parameters of a second order plant and compared to the MATLAB pidtune function. The GA method presented better results than MATLAB function for the most cases. The cases in that GA had a poor result were with high oscillations in unit step response of the plant. The explanation may the choice of the simple error function, thus a more sophisticated and complete error determination shall be studied. Although the advantage of the proposed GA method is that any prior knowledge of control system is not necessary to avoid complex analysis of high-order mathematical models.

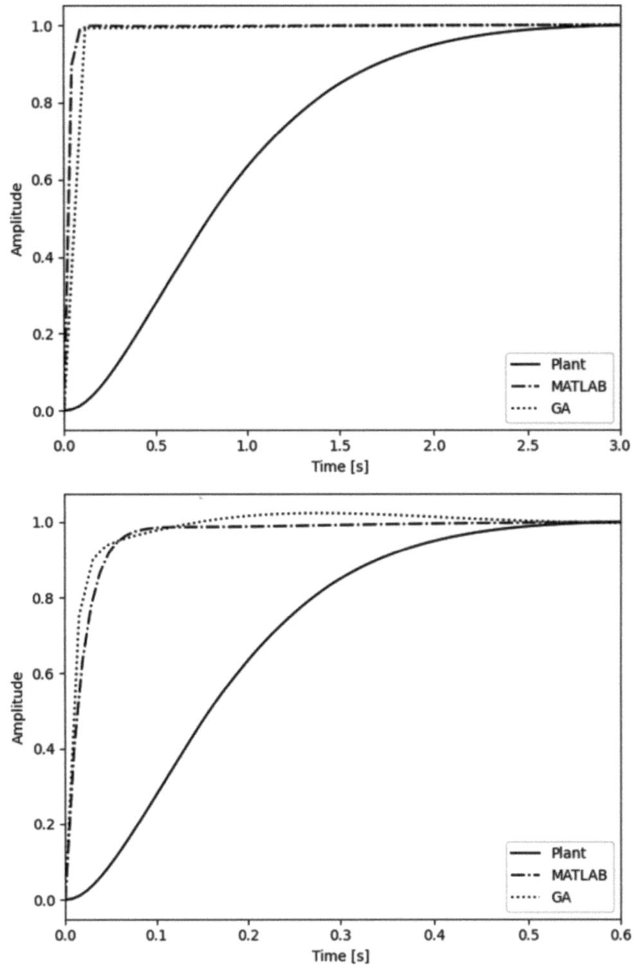

(a)

(b)

Figure 6.
Unit step response for $\xi = 0.9$ and (a) = 2 rad/s and (b) $\omega_n = 10$ rad/s.

Acknowledgements

The authors are grateful for the technical and financial support of UniRV (Rio Verde University), which made the publication of this book chapter possible. They also thank IntechOpen for the opportunity to share their studies and knowledge with other researchers.

Conflict of interest

The authors declare no conflict of interest.

Author details

Rodrigo de Figueiredo[1*], Bruno Toso[2] and Jean Schmith[2]

1 Rio Verde University, Rio Verde, Brazil

2 Unisinos University, São Leopoldo, Brazil

*Address all correspondence to: marquesf@unisinos.br

References

[1] Blondin MJ, Sanchis Sáez J, Pardalos PM. Control Engineering from Classical to Intelligent Control Theory—An Overview. In: Blondin M, Pardalos P, Sanchis Sáez J, editors. Computational Intelligence and Optimization Methods for Control Engineering. Springer Optimization and Its Applications. Vol. 150. Cham: Springer; 2019. DOI: 10.1007/978-3-030-25446-9_1

[2] Borase RP, Maghade D, Sondkar S, Pawar S. A review of pid control, tuning methods and applications. International Journal of Dynamics and Control. 2021; **9**(2):818-827

[3] Lloyds Raja G, Ali A. New pi-pd controller design strategy for industrial unstable and integrating processes with dead time and inverse response. Journal of Control, Automation and Electrical Systems. 2021;**32**(2):266-280

[4] Castellanos-C'ardenas D, Castrill'on F, V'asquez RE, Smith C. Pid tuning method based on imc for inverse-response second-order plus dead time processes. PRO. 2020;**8**(9):1183

[5] Nguyen NH, Nguyen PD. Overshoot and settling time assignment with pid for first-order and second-order systems. IET Control Theory & Applications. 2018;**12**(17):2407-2416

[6] Somefun OA, Akingbade K, Dahunsi F. The dilemma of pid tuning. Annual Reviews in Control. 2021;**52**:65-74

[7] Patel VV. Ziegler-Nichols tuning method. Resonance. 2020;**25**(10): 1385-1397

[8] Aisuwarya R, Hidayati Y. Implementation of Ziegler-Nichols PID Tuning Method on Stabilizing Temperature of Hot-water Dispenser. In: 2019 16th International Conference on Quality in Research (QIR): International Symposium on Electrical and Computer Engineering, Padang, Indonesia. 2019. pp. 1-5. DOI: 10.1109/QIR.2019.8898259

[9] Ekinci S, Hekimoğlu B, Kaya S. Tuning of PID Controller for AVR System Using Salp Swarm Algorithm. In: 2018 International Conference on Artificial Intelligence and Data Processing (IDAP), Malatya, Turkey. 2018. pp. 1-6. DOI: 10.1109/IDAP.2018.8620809

[10] Senthil Kumar S, Anitha G. A novel self-tuning fuzzy logic-based PID controllers for two-axis gimbal stabilization in a missile seeker. International Journal of Aerospace Engineering. 2021:1-12

[11] Anand A, Aryan P, Kumari N, Raja GL. Type-2 fuzzy-based branched controller tuned using arithmetic optimizer for load frequency control. Energy Sources, Part A: Recovery, Utilization, and Environmental Effects. 2022;**44**(2):4575-4596

[12] El-Samahy AA, Shamseldin MA. Brushless dc motor tracking control using self-tuning fuzzy pid control and model reference adaptive control. Ain Shams Engineering Journal. 2018;**9**(3): 341-352

[13] Somwanshi D, Bundele M, Kumar G, Parashar G. Comparison of fuzzy-pid and pid controller for speed control of dc motor using labview. Procedia Computer Science. 2019;**152**:252-260

[14] Potnuru D, Mary KA, Babu CS. Experimental implementation of flower pollination algorithm for speed controller of a bldc motor. Ain

Shams Engineering Journal. 2019;**10**(2): 287-295

[15] Bari S, Zehra Hamdani SS, Khan HU, Rehman Mu, Khan H. Artificial Neural Network Based Self-Tuned PID Controller for Flight Control of Quadcopter. In: 2019 International Conference on Engineering and Emerging Technologies (ICEET), Lahore, Pakistan. 2019. pp. 1-5. DOI: 10.1109/CEET1.2019.8711864

[16] Rodr'ıguez-Abreo O, Rodr'ıguez-Res'endiz J, Fuentes-Silva C, Hern'andez-Alvarado R, Falc'on MDCPT. Self-tuning neural network pid with dynamic response control. IEEE Access. 2021;**9**: 65206-65215

[17] Kumar D, Aryan P, Raja GL. Design of a novel fractionalorder internal model controller-based smith predictor for integrating processes with large dead-time. Asia-Pacific Journal of Chemical Engineering. 2022;**17**(1):e2724

[18] Kumari S, Aryan P, Raja GL. Design and simulation of a novel foimc-pd/p double-loop control structure for cstrs and bioreactors. International Journal of Chemical Reactor Engineering. 2021; **19**(12):1287-1303

[19] Mukherjee D, Raja G, Kundu P. Optimal fractional order imcbased series cascade control strategy with dead-time compensator for unstable processes. Journal of Control, Automation and Electrical Systems. 2021;**32**(1):30-41

Chapter 3

Perspective Chapter: Enhancing Network Efficiency Using Ant Colony Optimization and Pareto Optimality for Tuning PI Controller in Congested Routers

Samira Chebli

Abstract

The objective of this manuscript is to stabilize the queue of the router congestion window by designing an active queue management (AQM). The problem is dealt with under the theory of the command by using the regulator PI for that purpose. The tuning of this controller is based on a new approach of stabilization that relies on an extension of Hermite-Biehler theorem applied to quasi-polynomials. This stabilization method turns out relevant to seek the optimized results achieved within this stability region. For that, the optimization is performed using an improved multi-objective ant colony optimization (ACO) algorithm. The performance of the proposed control scheme is evaluated via a series of numerical simulations in MATLAB and Simulink.

Keywords: congestion control, AQM, PI controller, Hermite-Biehler theorem, ant colony optimization (ACO), time delay system, Pareto optimality, multi-objective

1. Introduction

The exponential increase in size and diversity of communication networks in the current era has become an issue for today's communities and a fertile field for researchers. In recent decades, researchers have been increasingly motivated to generate new and more reliable computer network structures that can handle the dimensions of current networks.

The TCP/IP protocol, a standard of network languages, is adopted on the Internet to facilitate communication between machines around the world. This universal language ensures that IP packets do not get lost or arrive in duplicate and confirm that the packet has arrived at its destination. However, as traffic increases, the network may become congested, a common phenomenon in routing. Routers may not be able to handle the traffic and lose packets because there is no place to store them. Queue saturation exacerbates congestion, as the time for a packet to reach the head of the queue becomes too long. In addition to queue saturation, these factors contribute to a slowdown of routers [1–5].

To quantitatively analyze the phenomenon of congestion, the behavior of the TCP protocol was modeled using a system of delayed differential equations based on the analogy of fluid flow. Previous research has shown that this system can be analyzed using the theory of control [6, 7].

Delay times resulting from congestion in Internet management give rise to characteristic functions known as quasi-polynomials. The first research on quasi-polynomials was conducted by Pontryagin [8], who developed a formal and highly relevant mathematical tool for analyzing the stability of time-invariant delayed systems. Pontryagin's necessary and sufficient condition for the roots of a quasi-polynomial to have a negative real part ensures Hurwitz stability for the quasi-polynomials in terms of a property of roots interlacing.

The issue with stabilizing delayed systems is that the characteristic equations of these models have an infinite number of roots, in contrast to systems without delay. The study of delayed systems presents a significant level of complexity, but a solution to this dilemma has been proposed in the work of [9–11]. This solution involves adopting an extension of the Hermite–Biehler theorem applied to quasi-polynomials, which is used to stabilize the TCP model using a proportional integrator (PI) controller [12, 13].

According to a survey conducted by the Japan Electric Measuring Instrument Manufacturers Association in 1989, the PI controller is used in more than 90% of industrial processes. This high usage is due to the unique characteristics of the PI controller, which provide stability, regardless of how optimal it may be. The proportional (P) and integral (I) actions of the PI controller complement each other to ensure immediate and rapid correction of any deviation of the quantity being adjusted. The PI controller eliminates large system inertias and residual steady-state error, while also accelerating system response and improving loop stability. Additionally, the PI controller can be combined with transmitters and logic, making it a strong candidate for use in the regulation of information transport, such as in TCP/IP networks.

The manuscript presents a problem that can be considered as a multi-objective optimization or Pareto optimality problem [14, 15]. The aim of the study is to minimize the rise time, settling time, and overshoot rate of the step response of the closed-loop system. However, these performance criteria are often conflicting, making simultaneous optimization challenging. To address this issue, the authors propose a multi-objective ant colony optimization (MOACO) technique based on the concept of Pareto optimality [16]. This technique guarantees reliable data transmission and is commonly used in routing problems. While previous studies have discussed ACO optimization for routing in TCP networks [17–19], this paper focuses on optimizing the flow of packets in a congested router by tuning the PI controller using the MOACO technique. Finally, the authors present a MATLAB simulation to demonstrate the performance of the proposed method (MOACO) for the PI-AQM control system.

2. Problem statement

2.1 The linearized fluid-flow model of TCP

Transmission control protocol/Internet protocol (TCP/IP) is the most widespread family of protocols that manage routing on the Internet. TCP provides a secure service to deliver packages as IP, it ensures the delivery of these packages.

The notoriety of the TCP/IP protocol is due to several factors including its reliability and its ability to allow a significant development of the Internet.

The reliability of transmission in TCP/IP networks is reflected in the use of acknowledgment packets (ACQs) and sequence numbers. When a node receives a packet, it sends an acknowledgment. A second acknowledgment for the same sequence number (duplicate acknowledgment) may be issued, meaning that a previous packet has been destroyed in the network or is delayed. This delay may be the result of increased traffic in the network, which may lead to the rejection of incoming frames in the buffers of the switches; this characterizes the phenomenon of congestion.

To address this issue, several active queue management (AQM) algorithms have been suggested, including random early detection (RED), adaptive RED (ARED) and Blue algorithms [20]. Such mechanisms which are basically heuristic methods tend to bring certain advantages such as drop tail enhancement by focusing on the length of the queue in order to reduce end-to-end delays and losses and, on the other hand, maximize the useful throughput in the network.

In this study, we focus on sharing a communication link between several senders as shown in **Figure 1**. It is assumed that the IP network has only one bottleneck, and that the link downstream of the router is borrowed by N flows.

The behavior of TCP underwent numerous studies, which have subsequently given rise to a mathematical representation of TCP analogous to fluid flow. This fluid modeling makes it possible to carry out a quantitative analysis of the congestion phenomenon. In this article, we consider the fluid model proposed by [6, 7]. The system is represented by a couple of stochastic differential equations taking into account the evolution of the congestion window but neglecting the mechanisms of slow start and time out. The linearization of the system model around a point of equilibrium [21] gives rise to the following expression:

$$G(s) = \frac{R^3C^3}{2\,N^2}\frac{e^{-Rs}}{\frac{R^3C}{N}s^2 + \left(R + \frac{R^2C}{N}\right)s + 2 + Rse^{-Rs}} \tag{1}$$

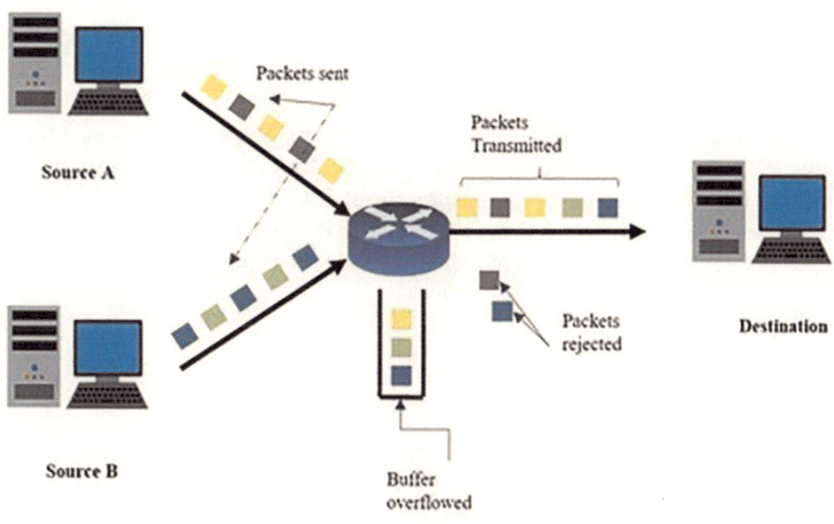

Figure 1.
Network topology studied.

2.2 The time-delay system model

Algorithms designed to regulate the length of queues may not always achieve their objectives, resulting in residual gaps. To compensate for these gaps, control theory approaches are employed. The aim of these approaches is to regulate the queue length to a specific threshold.

By reducing the congestion problem to a regulation problem, the system can be approximated as a first-order delay system, which is suitable for quantitative analysis without compromising the relevance of the results. This system model is described by

$$G(s) = \frac{Ke^{-Ls}}{Ts + 1} \tag{2}$$

Where

$K = \frac{R^3C^3}{4N^2}$ is the state gain,

$T = R\left(\frac{R^4C^2}{4N^2} + \frac{2RC}{N} - \frac{2R^2C}{N} + 2\right)^{\frac{1}{2}}$ is the constant time, and

$L = \frac{R^2C}{2N} + 2R - T$ is the time delay of the plant.

3. PI controller description

The focus of this section is on closed-loop stabilization using the PI controller for delayed linear invariant time systems (LTI).

The equation describing the PI regulator is given by

$$C(s) = K_p + \frac{K_i}{s} \tag{3}$$

Where K_p represents the proportional gain, and K_i represents the integral gain. The essential point of this study is the computation of the stability region defined by the parameters K_p and K_i.

The selection of the PI controller is based on its ability to minimize system error due to its various characteristics. Broadly speaking, the PI regulator achieves regulation by adjusting the system input based on the size of the error (which is the responsibility of the proportional or P action) and the duration for which the error has persisted (which is the responsibility of the integral or I action).

3.1 Hermite-Biehler theorem

This study utilizes an extension of the Hermite-Biehler theorem to quasi-polynomials that are characteristic equations of system models with delay, for stabilizing the system model with the PI regulator after modeling congestion as a first-order delay system. The challenge of studying this class of systems is due to the fact that the number of roots is infinite, unlike polynomials where the number of roots is finite. Early work in this field by Pontryagin [8] formulates a necessary and sufficient condition for the roots of a quasi-polynomial to have a negative real part. This work is later extended and enriched by Batcharaya, Silva, et al. [9–11, 13] for the study of quasi-polynomials.

This approach has been applied to congestion control and is elaborated upon in [21–23]. The approach allows for obtaining a complete set of parameters Kp and Ki, which ensures the stability of the closed-loop system, in contrast to conventional methods that rely on a single parameter. This is an essential step for finding optimal stabilizing parameters, which will be discussed in the following section.

3.2 Ant colony optimization

This section focuses on seeking optimal stabilization parameters K_p and K_i from the region of stability computed using the Hermite-Biehler theorem extension approach discussed in the previous section.

When faced with multiple paths between a starting point and an endpoint, the goal is to identify the most efficient route. Ants achieve this by exploring the different paths and leaving a trace (pheromone) along the way. Over time, the preferred paths are marked with more pheromone while less-used paths have their pheromones evaporate (as shown in **Figure 2**).

The ant colony optimization (ACO) approach, introduced by Marco Dorigo et al., is a population-based method inspired by the behavior of real ant colonies in finding food sources. This concept has been applied to complex combinatorial optimization problems that aim to identify optimal solutions within the constraints of the problem, such as congestion control in TCP networks [24, 25].

There are three essential phases in the ant colony algorithm:

1. Initialization: a set of ants is generated at the starting point.

2. Path construction: each ant selects a path to traverse based on the pheromone level and heuristic information.

3. Pheromone update: the amount of pheromone on the path is updated based on the quality of the solution found by the ant.

The ACO algorithm has been used to search for the optimal set of parameters K_p and K_i for the congestion control problem in TCP networks. In this case, the algorithm seeks to identify the best values for K_p and K_i that lead to stable performance within the constraints of the system [24, 25].

In the following, the three steps of ACO algorithm are elaborated.

Figure 2.
Ants converge to shortest path from their nest to the food.

3.2.1 Initialization

The objective is to identify the shortest Hamiltonian cycle within a graph, wherein each vertex of the graph symbolizes a distinct city. The distance between cities i and j is represented by d_{ij}, and the pair (i,j) represents the edge between these two cities. We first initialize the quantity of pheromone on edges with τ_{init}, and each ant traverses the graph and constructs a complete path (a solution).

3.2.2 Constructing ant solution

During each phase of the solution-building process, the ant must determine which location to move to, and this determination is made probabilistically based on the levels of pheromones present as well as statistical information. This approach enables the ant to discover a high-quality solution. The probability that an ant k moves from vertex i to vertex j, which belongs to a set of vertices that are not yet visited by the ant k denoted by S_i^k, is [26]:

$$P_{ij}^k = \frac{\left(\tau_{ij}(t)\right)^{\alpha}.\left(\eta_{ij}\right)^{\beta}}{\sum_{l \in S_i^k}(\tau_i(l))^{\alpha}.(\eta_{il})^{\beta}} \tag{4}$$

α and β are two parameters that influence the importance of the pheromone intensity τ_{ij}, and the statistical information called visibility η_{ij}. This value guides the choice of ants to nearby towns and avoids those that are too far away $\left(\eta_{ij} = \frac{1}{d_{ij}}\right)$. For $\alpha = 0$, one takes into account just the visibility, that is to say that the choice will have fallen each time on the nearest city. If $\beta = 0$, only the pheromone tracks play on the choice. To avoid too fast selection of a path, a suitable compromise between these two parameters is mandatory.

3.2.3 Updating pheromone

When all the ants have constructed a solution, an amount of pheromones $\Delta\tau_{ij}^k$ is deposited by each ant k on its path.

For any iteration t, if the path (i,j) is in the round of the ant k the quantity of pheromones deposited on this path is [27]:

$$\Delta\tau_{ij}^k(t) = \frac{Q}{L^k(t)} \tag{5}$$

Where $L^k(t)$ is the total length of the ant tour k, and Q is a constant.

The quantity of pheromones added is contingent upon the quality of the solution achieved, whereby lower pheromone amounts correspond to better solutions. To prevent disregard of inferior solutions and to prevent convergence toward local optima of inferior quality, a parameter is introduced to simulate the evaporation of pheromone trails. This parameter, ρ, called the evaporation rate $0 < \rho < 1$ is described as follows:

$$\tau_{ij}(t+1) \leftarrow (1-\rho)\tau_{ij}(t) + \Delta\tau_{ij}(t) \tag{6}$$

Where $\Delta\tau_{ij}(t) = \sum_{k=1}^{m} \Delta\tau_{ij}^{k}(t)$, t represents a given iteration and m the number of ants.

4. Design of PI controller using ACO with multiple objective optimization

In this section, we introduce the PI-MOACO controller, which employs an ACO algorithm to optimize the gains K_p and K_i used in congestion control by AQM, as demonstrated in **Figure 3**.

The ACO algorithm generates the gains K_p and K_i of the PI controller from the stability region. To leverage the ACO algorithm, it is preferable to represent the optimization problem using a direct path in the form of a construction graph, as illustrated in **Figure 4**.

The population is represented by a matrix (nx2), with the ant selecting the optimal parameters K_p and K_i of the PI controller by minimizing the objective function. **Figure 5** depicts the design problem utilizing the ant colony PI algorithm.

4.1 Cost functions used for the multi-objective ACO

Each parameter of K_p and K_i is encoded by n nodes in this study. Hence, a solitary node symbolizes the optimal values of K_p and K_i. The primary step in utilizing the

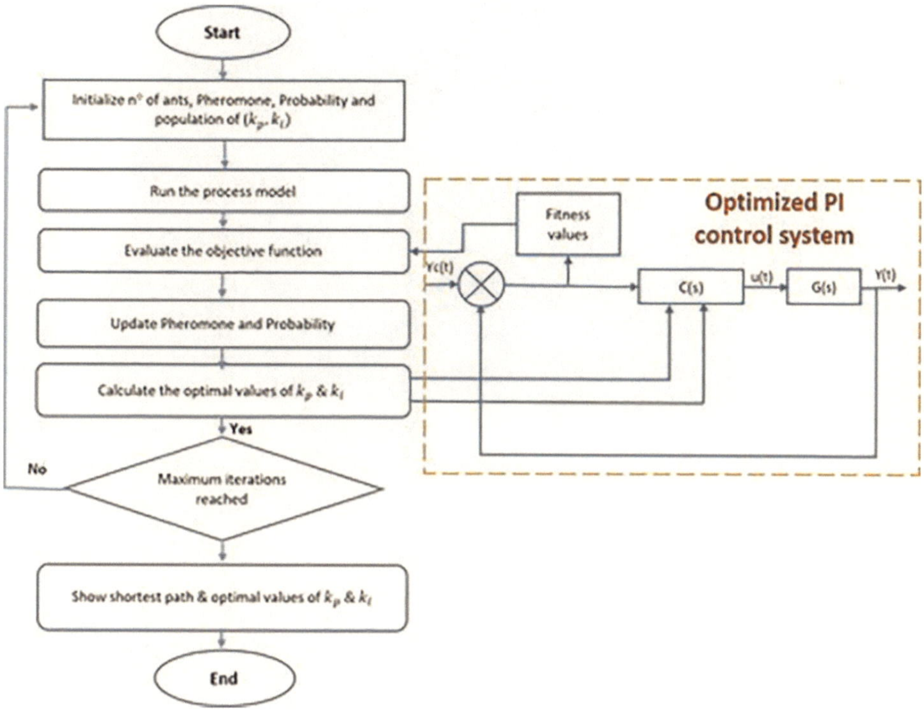

Figure 3.
Ant colony optimization flowchart for a PI proportional-integral controller.

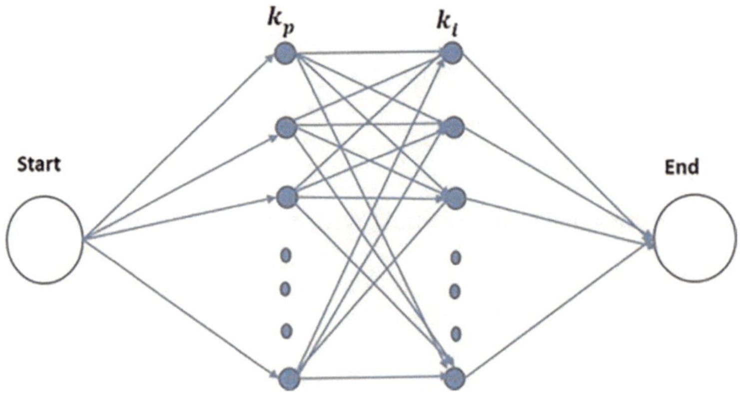

Figure 4.
ACO graph.

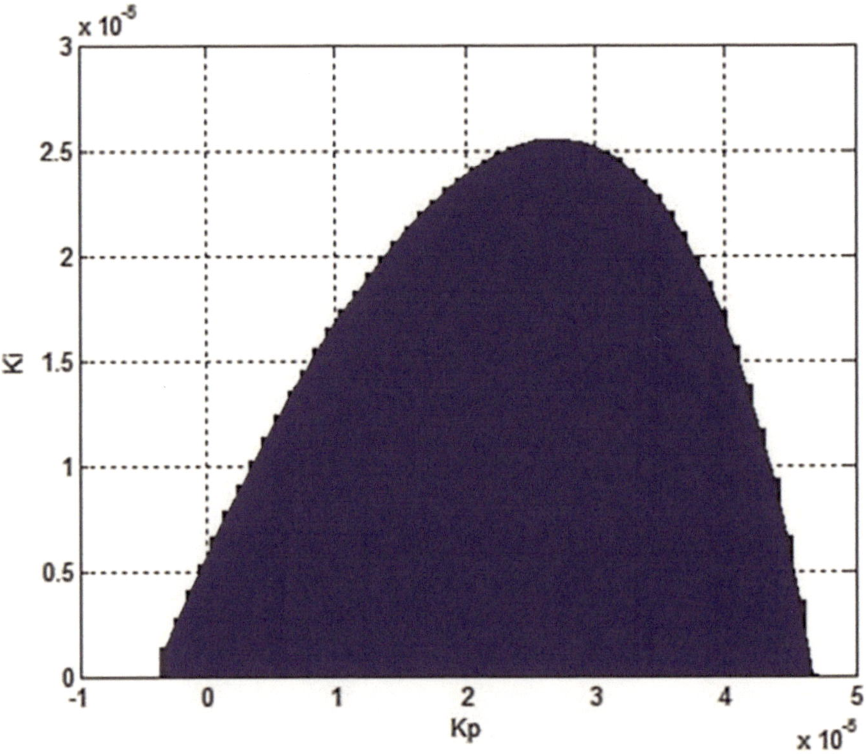

Figure 5.
Stabilizing region of (K_p, K_i) for the PI controller in the congestion control.

optimization technique is to select the criteria for optimization that assess fitness. Four performance indices, namely integrated-absolute error (IAE), integrated-squared error (ISE), integrated-time-absolute error (ITAE), and integrated-time-squared error (ITSE), are employed to minimize the error signal e(s). Their expressions are as follows:

$$\begin{cases} ISE = \sum_{0}^{t_{max}} e(t)^2 \\ IAE = \sum_{0}^{t_{max}} |e(t)| \\ ITAE = \sum_{0}^{t_{max}} t|e(t)| \\ ITSE = \sum_{0}^{t_{max}} te(t)^2 \end{cases} \tag{7}$$

where $e(t)$ is the error signal in time domain.

In the following, the ACO is characterized by a number of ants $m = 300$, evaporation rate $\rho = 0.7$, and the number of iterations = 50, $\alpha = 0.8$, and $\beta = 0.2$.

5. Simulation results and analysis

In this section, we assess the efficacy of the closed-loop system using the MOACO with PI controller via simulation. The simulation is performed using MATLAB, with an example from the literature model system used to illustrate the study presented throughout the paper, with the aim of optimally stabilizing the system.

$$\hat{G}(s) = \frac{2.34.10^5 e^{-0.77s}}{5.03s + 1} \tag{8}$$

By applying the Hermite–Biehler extension approach, we find that the set of K_p stabilizing values is given below:

$$K_p \in \left[-4.27.10^{-6}; 3.89.10^{-5}\right]. \tag{9}$$

Figure 5 provides a two-dimensional representation of the complete set of stabilizing PI controller variables K_p and k_i for the TCP/AQM system. The MOACO method provides the PI controller's optimum parameters, which are given in **Table 1**. **Figure 6** presents the step response for the different PI controllers optimized by MOACO.

To analyze the dynamic performance of the system studied, we consider three performance criteria, namely settling time, rise time, and overshoot rate, among others.

Performances criteria	K_p (10^{-5})	K_i (10^{-5})
IAE	1.85	1.08
ISE	1.60	1.02
ITAE	2.1	1.11
ITSE	2.35	1.12

Table 1.
Optimum parameters of PI controller.

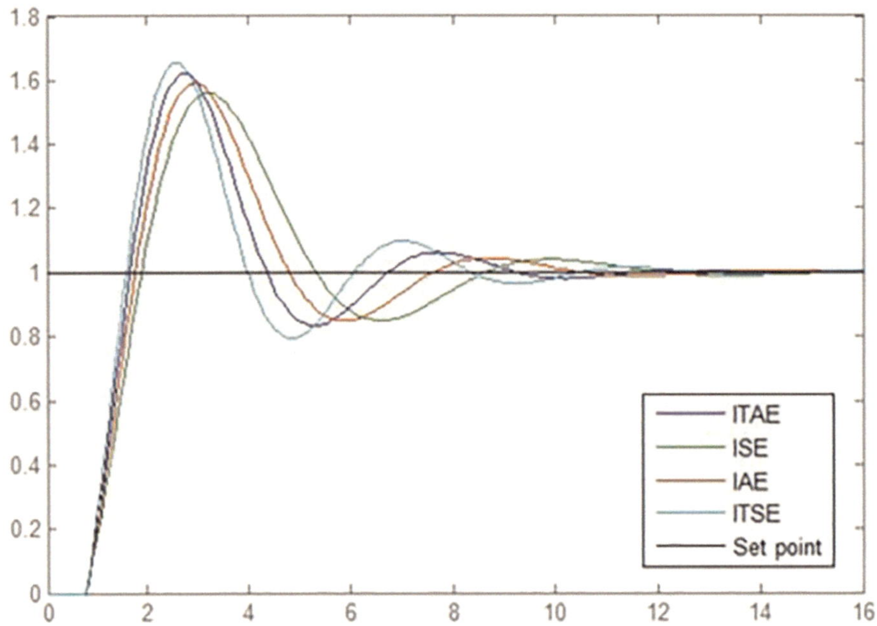

Figure 6.
Step response by different objectives with MOACO-PI controller.

Settling time is the time taken to reach the final value of the output to within 5%.

Rise time is the time at which the output signal crosses its asymptote for the first time. The rise time is a relevant parameter for encrypting the speed of a closed-loop system, which is the ultimate objective of our study.

Overshoot rate measures how much the output exceeds the steady-state value before settling.

Table 2 summarizes these standard performance measures.

The results show a conflict between the performance indices, which is expected since we use an approach based on the Pareto optimality principle. From **Table 2**, we can see that ITSE provides the best rise time among all the other objectives. The best value of settling time is given by the IAE criterion, and the smallest overshoot rate is provided by the ISE cost function. It has been highlighted that the main purpose of this study is to minimize the time delay occurring through the TCP network, with guaranteed stability. From the table above, we can conclude that these objectives are achieved since we obtain quite small rise and settling times.

Performances criteria	Rise time (s)	Settling time (s)	Overshoot (%)
IAE	0.79	9.80	59.15
ISE	0.89	11.14	56.17
ITAE	0.72	10.31	62.15
ITSE	0.66	9.96	65.58

Table 2.
Numerical values of standard performance measures.

6. Conclusion

This paper addresses the problem of congestion, one of the most common challenges encountered in communication networks. The proposed approach involves computing the stability region for a first-order delay system controlled by a PI controller using an extension of the Hermite-Biehler theorem applied to quasi-polynomials. The next step involves using a multi-objective optimization method based on the ant colony optimization (ACO) algorithm to search for optimal PI controller gains K_p and K_i within the stability region. Four performance criteria calculated in closed-loop plant are used as objective functions. Simulation results demonstrate the effectiveness of the proposed method in terms of dynamic performance, including reduction of maximum overshoot, rise time, and settling time. The MOACO algorithm allows for rapid convergence in local research, highlighting the flexibility and efficiency of the proposed approach.

Overall, the simulation results support the effectiveness of the proposed approach in addressing the problem of congestion in communication networks.

Funding

This research received no external funding.

Conflict of interest

The author declares no conflict of interest.

Author details

Samira Chebli
Control, Management and Monitoring of Electrical Energy's Lab, Mines School of Rabat, Rabat, Morocco

*Address all correspondence to: samira.chebli@gmail.com

References

[1] Jacobson V. Congestion avoidance and control. ACM SIGCOMM Computer Communication Review. 1995;**25**(1): 157-187

[2] Chiu D, Jain R. Analysis of the increase/decrease algorithms for congestion avoidance in computer networks. Journal of Computer Networks and ISDN. 1989;**17**(1): 1-14

[3] Ohsaki H, Sugiyama K, Imase M. Congestion propagation among routers with TCP flows, international. Journal of Computer Networks & Communications. 2009;**1**(2):112-127

[4] Alshimaa HI, el Sayed A, Elsaghir Z, Morsi IZ. Enhanced random early detection (ENRED). International Journal of Computer Applications. 2014; **92**:9

[5] Low HS, Paganini F, Doyle JC. Internet congestion control. IEEE Control Systems Magazine. 2002;**22**: 28-43

[6] Misra V, Gong W, Towsley D. Stochastic differential equation modeling and analysis of TCP windowsize behavior. In: Proceedings of PERFORMANCE. Vol. 99. USA: University of Massachusetts; 1999

[7] Misra V, Gong W, Towsley D. Fluid-based analysis of a network of AQM routers supporting TCP flows with an application to RED. In: Proceedings of the Conference on Applications, Technologies, Architectures, and Protocols for Computer Communication. Stockholm, Sweden. 2000. pp. 151-160

[8] Pontryagin LS. On the zeros of some elementary transcendental function. American Mathematical Society Translation. 1955;**2**:95-110

[9] Silva GJ, Datta A, Bhattacharyya SP. New results on the synthesis of PID controller. IEEE Transactions on Automatic Control. 2002;**47**(2)

[10] Silva GJ, Datta A, Bhattacharyya SP. Stabilization of time delay systems. In: Proceedings of the American Control Conference. Chicago, IL, USA. 2000. pp. 963-970

[11] Silva GJ, Datta A, Bhattacharyya SP. PID Controllers for Time Delay Systems. London: Springer; 2005

[12] Silva GJ, Datta A, Bhattacharyya SP. Stabilization of first-order systems with time delay using the PID controller. In: Proceedings of the American Control Conference. Arlington, VA, USA. 2001. pp. 4650-4655

[13] Bhattacharyya SP, Datta A, Keel LH. Linear Control Theory: Structure, Robustness and Optimization. USA: CRC Press, Taylor & Francis; 2009

[14] Chinchuluun A et al. Pareto optimality, game theory and equilibria. In: Springer Optimization and Its Applications. Berlin, Germany. 2008

[15] Censor Y. Pareto optimality in multiobjective problems. Applied Mathematics and Optimization. 1977; **4**(1):41-59s. DOI: 10.1007/BF01442131

[16] Zheng Z et al. Meta-heuristic techniques in microgrid management: A survey. Swarm and Evolutionary Computation. 2023;**2023**:101256

[17] Rathore S, Khan MR. Performance of TCP and UDP protocols for secure

multipath aco communication in manet. International Journal of Development Research. 2017;**07**(01):11214-11218

[18] Ring S et al. Ant colony optimization based model for network zero-configuration. In: Signal Processing and Communications International Conference, Bangalore, India. 2004. DOI: 10.1109/SPCOM.2004.1458494

[19] Ramamoorthy R, Thangavelu M. An enhanced distance and residual energy-based congestion aware ant colony optimization routing for vehicular ad hoc networks. International Journal of Communication Systems. 2022;**35**(11): e5179

[20] Giménez A et al. New RED-type TCP-AQM algorithms based on beta distribution drop functions. Applied Sciences. 2022;**12**(21):11176

[21] Chebli S, Elakkary A, Sefiani N, Elalami N. New PI stabilization of first-order congestion control of active queue management routers. In: Proceeding of Electrical and Information Technologies (ICEIT) Conference; March 2015; Marrakech, Morocco. pp. 12-217

[22] Chebli S. PI stabilization for congestion control of AQM routers with tuning parameter optimization. International Journal of Interactive Multimedia and Artificial Intelligence. 2016;**4**(1):52-55

[23] Chebli S, Elakkary A, Sefiani N. Design of an Optimal PI (Proportional–Integral) Controller for the Robust Control of Uncertain TCP Traffic. USA: NOVA Science Publisher; 2021. pp. 39-106

[24] Dorigo M. Optimization, Learning and Natural Algorithms (in Italian). Italy: Dipartimento di Elettronica, Politecnico di Milano; 1992

[25] Dorigo M, Maniezzo V, Colorni A. The ant system: Optimization by a colony of cooperating agents. IEEE Transactions on Systems, Man, and Cybernetics – Part B. 1996;**26**(1): 29-41

[26] Abolhasan M, Wysocki T, Dutkiewicz E. A review of routing protocols for mobile ad hoc networks. Ad Hoc Networks. 2004;**2**(1):1-22

[27] Bullnheimer B, Hartl RF, Strauss C. An improved ant system algorithm for the vehicle routing problem. Annals of Operations Research. 1999;**89**:319-328

Chapter 4

Observer-Based Disturbance Rejection Control for Switched Nonlinear Networked Systems under Event-Triggered Scheme

Arumugam Arunkumar and Jeng-Lang Wu

Abstract

This paper employs the disturbance rejection technique for a class of switched nonlinear networked control systems (SNNCSs) with an observer-based event-triggered scheme. To estimate the influence of exogenous disturbances on the proposed system, the equivalent input disturbance (EID) technique is employed to construct an EID estimator. To provide adequate disturbance rejection performance, a new control law is built that includes the EID estimation. Furthermore, to preserve communication resources, an event-based mechanism for control signal transmission is devised and implemented. The primary goal of this work is to provide an observer-based event-triggered disturbance rejection controller that ensures the resulting closed-loop form of the examined systems is exponentially stable. Specifically, by employing a Lyapunov–Krasovskii approach, a new set of sufficient conditions in the form of linear matrix inequalities (LMIs) is derived, ensuring the exponential stabilization criteria are met. Eventually, a numerical example is used to demonstrate the efficacy and practicality of the proposed control mechanism.

Keywords: nonlinear networked systems, disturbance rejection control, event-triggered scheme, equivalent input disturbance, Lyapunov techniques

1. Introduction

The switched system is a very flexible modeling tool consisting of a family of distinct subsystems, operated according to a specific switching rule, that determines which subsystem is active at any given time [1, 2]. Switched systems are employed in a variety of real-world settings, including power systems, networked control systems, communication networks, electrical devices and circuits, underwater vehicle systems, manipulator robots, and many more areas [3, 4]. In recent years, a considerable amount of research has been conducted on the stability and stabilization of switched systems using Lyapunov techniques. For instance, the existence of a common Lyapunov function for individual systems that ensures stability of the switched system in the context of arbitrary switching sequence, see [5, 6]. The authors in [7] discussed the stability and stabilization problem of switched linear systems along with average dwell time approach, where as in [4], the issue of switched nonlinear systems under multiple time delays is discussed.

On the other hand, networked control systems (NCSs) have received considerable attention due to their advantages over traditional point-to-point wired control techniques, including convenience extension, information sharing, affordability, ease of installation and maintenance, and high reliability [8]. As a result, they have recently been widely used in many different fields, such as robotic manipulators, vehicle highway systems, spacecraft systems, teleoperation systems, smart grids, and other areas. It is common for NCSs to use wired and wireless communications networks to link the various components of the system, including sensors, controllers, and actuators, which are situated in separate locations, in order to form a feedback control system [5, 9]. During the past few years, switched NCS's have received a great deal of attention from researchers [10, 11] due to the fact that these kind of systems can well describe numerous practical systems with abrupt parameter variations. However, the communication network introduces some new difficulties to the controller design of switched NCSs [12].

Furthermore, some interesting issues may arise due to the characteristics of the network, such as congestion on the network, data loss, and disorder, and delay due to the network. In order to reduce the network resources, the time-triggered technique has been established in several versions. Over a specified time period, every data is sampled and transmitted, whether required or not. In the case that if the difference between the latest transmitted signal and the current sampling signal is relatively small, then a significant amount of the same data will be transmitted in the network, which will cause unnecessary traffic on the network and unwanted communication resources. To tackle these issues, quite of few researchers devote themselves in developing extremely effective data transmission techniques, for example, [13, 14]. Motivated by [5], an event-triggered technique has been developed to further reduce the limited bandwidth and enhance the usage of communication resources. The authors in [10] propose an event-triggered H_∞ communication control problem for switched networked systems with the assistance of the LKF theory. Further, the authors in [9] discussed an event-triggered H_∞ control strategy for NCSs under communication delay. Under an adaptive event-triggered communication scheme framework, an H_∞ tracking control problem for nonlinear networked systems is derived with limited network communication [15]. In particular, the distributed H_∞ control problem for switched networked linear control system is discussed in [16] subject to quantization and packet dropouts. In [3], an observer-based controller design has been developed for the switched networked nonlinear systems with packet dropouts and average dwell-time mechanisms.

As a result, it is more important from the perspective of the control system to evaluate exogenous disturbances affecting the control input channel since they are a source of not only the collapse of the system's control performance but also machine malfunctions, vibrations, and other issues. In order to achieve good disturbance rejection performance, the EID method, such as the one described in [17, 18], is an active disturbance rejection technique with two degrees of freedom, which actively processes disturbance information and constructs compensation signals for them.

The main advantage of employing the EID technique is that no restrictions are imposed or demand any information concerning disturbances [2]. As a general rule, EID is capable of rejecting and estimating both matched and mismatched external disturbances, and its configuration allows for easy modeling. Several design methodologies have been proposed in order to take into account the energetic performance of EID, as recently examined in the literature, see, for example, [19, 20]. Based on the disturbance rejection method, the switched neutral time-delay system is relaxed in [2] via EID approach. The authors in [18] derived an effective disturbance rejection method for time delay system based on the EID approach. She et al. [21] explored the problem of enhancing the disturbance-rejection performance for servo systems, which involves the estimate of an EID and an improved servo system design using the EID technique.

Moreover, it has been reported that innumerable advanced techniques on an event-triggered control of switched networked systems have already been published, but as far as we are aware, this has not yet been explored for an observer-based event-triggered control scheme for SNNCSs, especially for disturbance rejection technique and ADT approach. As stipulated by the author's perspective, the main intention of this paper is to investigate the EID method to reject disturbances for SNNCSs in the presence of observer-based event-triggered control mechanism. Roughly stated, the novelty of this paper is contributed as follows:

- An unified disturbance rejection and observer-based control design for SNNCSs with external disturbances and an event-triggered scheme is developed in this study.

- One aspect of the proposed event-triggered DRC scheme is that in contrast to the traditional time-triggered-DRC scheme [22, 23], it can significantly reduce communication frequency while achieving satisfactory closed-loop system performance by utilizing a new event-triggered transmission scheme. However, many existing event-triggered approaches without direct disturbance compensation have a tendency to violate the event triggering conditions when disturbances occur [15], while the ET-ADRC can be effectively used to reduce the occurrence times of triggering conditions by utilizing active disturbance compensation.

- The proposed controller includes an EID estimator to efficiently estimate and reduce external disturbances in the system understudy.

- Furthermore, by resorting to a Lyapunov–Krasovskii functional, adequate constraints are attained in the context of LMIs. It follows that the requisite controller gain matrices are determined by solving the established conditions.

Finally, to show the utility of the proposed control scheme, a realistic example is given.

2. Problem description

In this article, we focus on obtaining an observer-based event-triggered control strategy for SNNCSs with disturbance and achieving disturbance rejection. To save the limited network resources, an event-triggered generator is employed.

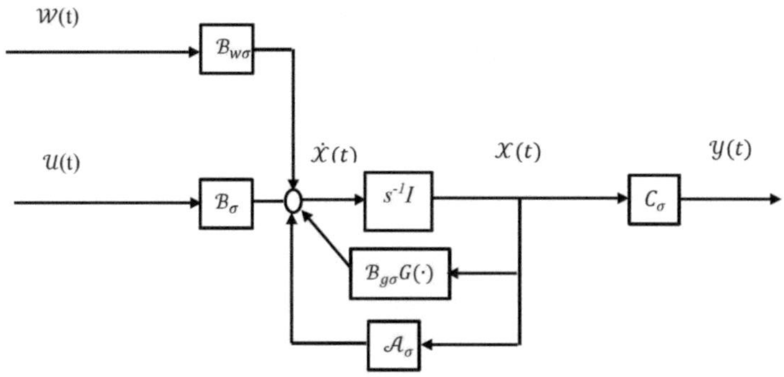

Figure 1.
Controlled plant.

2.1 Controlled plant

As illustrated in **Figure 1**, the controlled plant is modeled by the following SNNCSs with external disturbance:

$$\begin{aligned} \dot{\mathcal{X}}(t) &= \mathcal{A}_{\sigma(t)}\mathcal{X}(t) + \mathcal{B}_{\sigma(t)}\mathcal{U}(t) + \mathcal{B}_{\mathcal{G}\sigma(t)}\mathcal{G}(\mathcal{X}(t)) + \mathcal{B}_{\mathcal{W}\sigma(t)}\mathcal{W}(t), \\ \mathcal{Y}(t) &= \mathcal{C}_{\sigma(t)}\mathcal{X}(t), \\ \mathcal{Z}(t) &= \mathcal{E}_{\sigma(t)}\mathcal{X}(t) + \mathcal{F}_{\sigma(t)}\mathcal{W}(t), \end{aligned} \tag{1}$$

where $\mathcal{X}(t) \in \mathbb{R}^{\mathcal{N}_x}$ is the state vector; $\mathcal{Y}(t) \in \mathbb{R}^{\mathcal{N}_y}$ and $\mathcal{Z}(t) \in \mathbb{R}^{\mathcal{N}_z}$ are the measured and the controlled outputs of the system; $\mathcal{W}(t) \in \mathbb{R}^{\mathcal{N}_w}$ is the disturbance input signal, which belongs to $\mathcal{L}_2[0, \infty)$; $\mathcal{U}(t) \in \mathbb{R}^{\mathcal{N}_u}$ is the ideal control signal generated by the controller; $\sigma(t)$ represents the switching signal that accepts values from a finite collection $\mathcal{I} = \{1, 2, \cdots, \mathcal{R}\}$; for example, when $\sigma(t) = i \in \mathcal{I}$, it implies that the ith subsystem has been enabled, where $\mathcal{R}$ denotes the number of subsystems; The matrices $\mathcal{A}_i, \mathcal{B}_i, \mathcal{B}_{\mathcal{G}i}, \mathcal{B}_{\mathcal{W}i,}\mathcal{C}_i, \mathcal{E}_i$ and $\mathcal{F}_i, i = 1, \cdots, \mathcal{R}$, are known real constant matrices with appropriate dimensions; furthermore, $\mathcal{G}(\mathcal{X}) = [\mathcal{G}_1(\mathcal{X}), \mathcal{G}_2(\mathcal{X}), \ldots \mathcal{G}_n(\mathcal{X})]^T$ is a nonlinear vector function, which fulfills the global Lipschitz requirement $\|\mathcal{G}(t, \mathcal{X}) - \mathcal{G}(t, \hat{\mathcal{X}})\| \leq \beta_{\mathcal{G}}\|\mathcal{X} - \hat{\mathcal{X}}\|, \forall \mathcal{X}, \hat{\mathcal{X}} \in \mathbb{R}^{\mathcal{N}_x}$, where $\beta_{\mathcal{G}}$ is a positive scalar.

In order to use the EID-based technique to analyze SNNCSs, we must first examine the EID's definition.

Definition 1 Let the input $\mathcal{U}(t)$ in the SNNCSs plant Eq. (1) be zero. A signal, $\mathcal{W}_e(t)$, on the control input channel is called an EID of the disturbance $\mathcal{W}(t)$, if it produces the same impact on the output as the disturbance $\mathcal{W}(t)$ does for all $t \geq 0$.

2.2 Configuration of the EID-based SNNCSs

For the purpose of enhancing disturbance-rejection performance for the SNNCSs, Liu et al. [18] suggest an EID technique. The EID-based control system setup has two degrees of freedom. As a consequence, it makes it possible for this technique to actively examine the information from the disturbance and develop a compensation for it. A disturbance estimator is crucial in this procedure for compensating the disturbances. It has a significant impact on disturbance rejection.

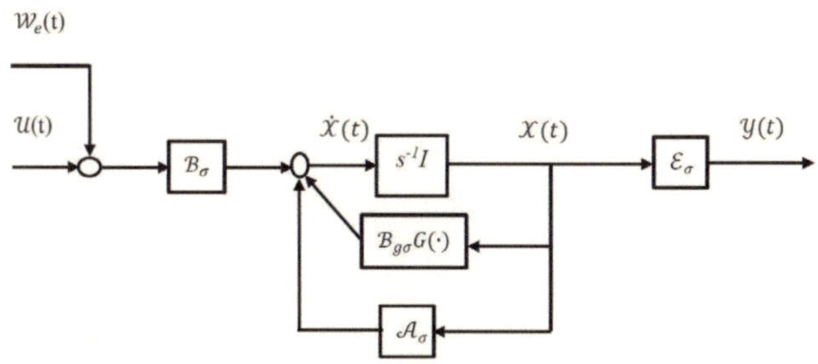

Figure 2.
Controlled plant with EID.

It should be noted that since $\mathcal{B}_{\sigma(t)}$ and $\mathcal{B}_{\mathcal{W}\sigma(t)}$ may have various dimensions, the disturbance may be imposed on a channel other than the control input channel and that the number of disturbances and related input channels may also be greater than one. However, if we suppose that a disturbance is exclusively placed on the control input channel, as illustrated in **Figure 2**, then the plant in terms of Definition 1 as

$$\dot{\mathcal{X}}(t) = \mathcal{A}_{\sigma(t)}\mathcal{X}(t) + \mathcal{B}_{\mathcal{G}\sigma(t)}\mathcal{G}(\mathcal{X}(t)) + \mathcal{B}_{\sigma(t)}(\mathcal{U}(t) + \mathcal{W}_e(t)). \tag{2}$$

In practical implementation, it should be noted that the entire states of the physical plant Eq. (2) are not completely measurable. Taking this fact into account, an observer-based controller can be constructed by means of measurement output to estimate the states. The state observer is:

$$\begin{aligned} \dot{\hat{\mathcal{X}}}(t) &= \mathcal{A}_{\sigma(t)}\hat{\mathcal{X}}(t) + \mathcal{B}_{\sigma(t)}\mathcal{U}_{\mathcal{F}}(t) + \mathcal{B}_{\mathcal{G}\sigma(t)}\mathcal{G}(\hat{\mathcal{X}}(t)) + \mathcal{L}_{\sigma(t)}(\mathcal{Y}(t) - \hat{\mathcal{Y}}(t)), \\ \hat{\mathcal{Y}}(t) &= \mathcal{C}_{\sigma(t)}\hat{\mathcal{X}}(t), \end{aligned} \tag{3}$$

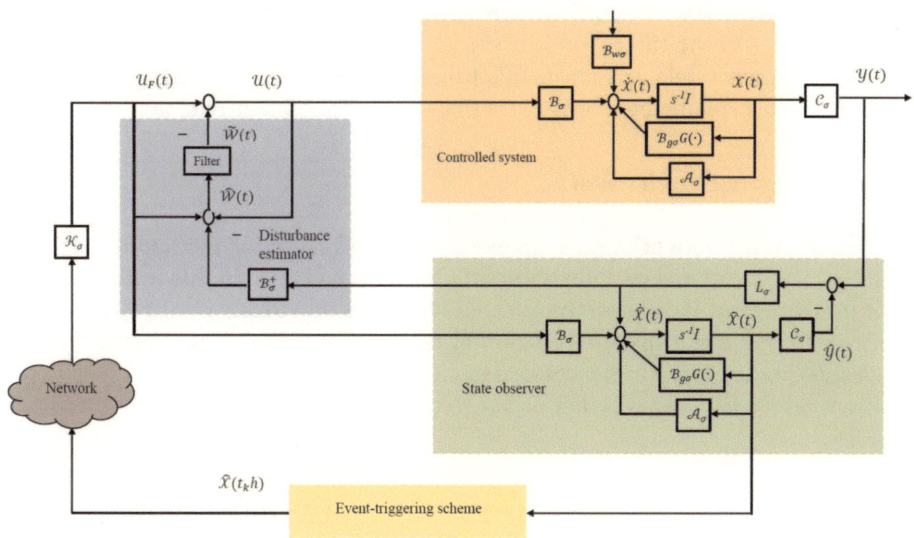

Figure 3.
Controlled plant with EID-based event-triggered mechanism.

where $\hat{\mathcal{X}}(t) \in \mathbb{R}^{\mathcal{N}_x}$ and $\hat{\mathcal{Y}}(t) \in \mathbb{R}^{\mathcal{N}_y}$ are the reconstructed state and output vectors of $\mathcal{X}(t)$ and $\mathcal{Y}(t)$, respectively; $\mathcal{U}_{\mathcal{F}}(t)$ is the control input vector of the system, introduced below; and $\mathcal{L}_i, i = 1, \cdots, \mathcal{R}$, are the observer gain matrix to be determined.

The observer-based feedback control law $\mathcal{U}_{\mathcal{F}}(t)$, shown in **Figure 3** is designed as follows:

$$\mathcal{U}_{\mathcal{F}}(t) = \mathcal{K}_{\sigma(t)}\hat{\mathcal{X}}(t) \tag{4}$$

where $\mathcal{K}_i, i = 1, \cdots, \mathcal{R}$ denotes the control gain matrices.

It should be emphasized that SNNCS Eq. (1) may contain disturbances, which could result in the system's poor performance or instability. In order to estimate and reject both matched and unmatched disturbances, the EID technique does not require a priori knowledge of the disturbances. In this SNNCSs, an EID technique is employed in the control channel, which yields satisfactory disturbance rejection performance. According to discussed in [17, 18], the optimal EID estimated disturbance can be represented by

$$\begin{aligned} \hat{\mathcal{W}}(t) &= \mathcal{B}^{+}_{\sigma(t)}\mathcal{L}_{\sigma(t)}\mathcal{C}_{\sigma(t)}\left(\mathcal{X}(t) - \hat{\mathcal{X}}(t)\right) + \mathcal{U}_{\mathcal{F}}(t) - \mathcal{U}(t), \\ \text{where} \quad \mathcal{B}^{+}_{\sigma(t)} &= \left(\mathcal{B}^{T}_{\sigma(t)}\mathcal{B}_{\sigma(t)}\right)^{-1}\mathcal{B}^{T}_{\sigma(t)}. \end{aligned} \tag{5}$$

In order to filter the measurement noise in the estimated disturbance, let $\hat{\mathcal{W}}(t)$ pass through a filter represented as

$$\begin{aligned} \dot{\mathcal{X}}_{\mathcal{F}}(t) &= \mathcal{A}_{\mathcal{F}\sigma(t)}\mathcal{X}_{\mathcal{F}}(t) + \mathcal{B}_{\mathcal{F}\sigma(t)}\hat{\mathcal{W}}(t), \\ \tilde{\mathcal{W}}(t) &= \mathcal{C}_{\mathcal{F}\sigma(t)}\mathcal{X}_{\mathcal{F}}(t) \end{aligned} \tag{6}$$

where $\mathcal{X}_{\mathcal{F}}(t) \in \mathbb{R}^{\mathcal{N}_{x_{\mathcal{F}}}}$ is the state of the filter $\mathcal{F}(s)$, whereas $\tilde{\mathcal{W}}(t)$ represents the output of the filter, and $\mathcal{A}_{\mathcal{F}\sigma(t)}, \mathcal{B}_{\mathcal{F}\sigma(t)}$ and $\mathcal{C}_{\mathcal{F}\sigma(t)}$ are known parameters. Moreover, the transfer function of the filter $\mathcal{F}(s)$ satisfies $|\mathcal{F}(j\omega)| \approx 1, \forall \omega \in [0, \omega_r]$, where ω_r is the maximum frequency selected for the disturbance estimation [18]. An appropriate filter has a cutoff frequency that is more than ten times greater than ω_r.

2.3 Event-triggered mechanism

To decrease the network usage, an event-triggered sampling method is used in place of the traditional periodic sampling technique [1, 11]. In this technique, sensor measurement is only communicated when the previously transmitted value and the value of a specific function of the current sensor measurement reach a threshold value [5]. The event detector considers the following conditions while deciding whether or not to transmit the current signal to the controller:

$$t_{k+1}h = t_k h + \min_{s}\left\{sh \,|\, \hat{e}^T(t_k h + sh)\Psi_{1\sigma(t)}\hat{e}(t_k h + sh) > \varsigma_{\sigma(t)}\hat{x}^T(t_k h)\Psi_{2\sigma(t)}\hat{x}(t_k h)\right\}, \tag{7}$$

where $0 < \varsigma_i < 1$, $i = 1, 2, \cdots, \mathcal{R}$ are known parameters, $\Psi_{1\sigma(t)}$ and $\Psi_{2\sigma(t)}$ are symmetric positive definite matrices to be constructed, $h > 0$ is the sampling period, and

$\hat{e}(t_k h + sh)$ is the difference between the two state estimations at the last transmission instant and the present sampling instant, that is, $\hat{e}(t_k h + sh) = \hat{\mathcal{X}}(t_k h) - \hat{\mathcal{X}}(t_k h + sh)$.

When the data provided by the event monitor is transferred to the controller at t_k, a communication delay known as the sensor-to-controller delay $\tau_{sc}(t_k)$ is produced. Similarly, the controller's transmission of actuation signals to the actuator at t_k causes another communication delay known as the controller-to-actuator delay $\tau_{ca}(t_k)$. The sum of these two delays is represented by $\tau_{t_k} = \tau_{sc}(t_k) + \tau_{ca}(t_k)$ with $0 < \tau_{t_k} < \tau_M$, where τ_M represents the maximum delay bounds of τ_{t_k} [9]. Finally, the actuator actuates at the time specified by $t_k h + \tau_{t_k}$. During the network transmission, the input of the controlled plant at the time interval $\left[t_k h + \tau_{t_k}, t_{k+1} h + \tau_{t_{k+1}}\right)$ is denoted by

$$\hat{\mathcal{U}}_{\mathcal{F}}(t) = \mathcal{K}_{\sigma(t)} \hat{\mathcal{X}}(t_k h) \tag{8}$$

We decompose the holding interval of zero-order-hold (ZOH) $\left[t_k h + \tau_{t_k}, t_{k+1} h + \tau_{t_{k+1}}\right)$ into the following subintervals as in [24]: $\left[t_k h + \tau_{t_k}, t_{k+1} h + \tau_{t_{k+1}}\right) = \cup_{s=0}^{t_{k+1}-t_k-1} T_s$, where $T_s = \left[i_k h + \tau_{i_k}, i_k h + h + \tau_{i_k+1}\right)$ and $i_k h = t_k h + sh, s = 0, 1, \cdots, t_{k+1} - t_k - 1$ signify the sampling instants that occur between the present sampling instant $t_k h$ and the future sampling instant $t_{k+1} h$. If s takes the value of $t_{k+1} - t_k - 1$, then $\tau_{i_k+1} = \tau_{t_k+1}$, otherwise $\tau_{i_k} = \tau_{t_k}$. The network permissible equivalent delay $\tau(t)$ is now defined as $\tau(t) = t - i_k h,\ \ t \in T_s$, then we have $0 \le \tau(t) \le \tau_M$.

2.4 Analysis and design of the closed-loop SNNCSs

Eventually, the filter estimated disturbance $\tilde{\mathcal{W}}(t)$ together with observer-based control law $\mathcal{U}_{\mathcal{F}}(t)$ yields a new controller, which is given by

$$\mathcal{U}(t) = \hat{\mathcal{U}}_{\mathcal{F}}(t) - \tilde{\mathcal{W}}(t) = \mathcal{K}_{\sigma(t)} \hat{\mathcal{X}}(t_k h) - \tilde{\mathcal{W}}(t) \tag{9}$$

The system internal stability condition does not depend on exogenous signals, hence the exogenous signals $\mathcal{W}(t)$ assumed to be zero when analyzing the internal stability. Furthermore, we define the error between the state Eq. (1) and the observer Eq. (3) with the configuration $\Delta\mathcal{X}(t) = \mathcal{X}(t) - \hat{\mathcal{X}}(t)$. We use the states of the control system $\left(\hat{\mathcal{X}}(t), \Delta\mathcal{X}(t), \mathcal{X}_{\mathcal{F}}(t)\right)$ to define $\varphi(t) = \left[\hat{\mathcal{X}}(t) \quad \Delta\mathcal{X}(t) \quad \mathcal{X}_{\mathcal{F}}(t)\right]^T$ and use it to describe the closed-loop system. Since

$$\begin{aligned}
\dot{\hat{\mathcal{X}}}(t) &= \mathcal{A}_{\sigma(t)} \hat{\mathcal{X}}(t) + \mathcal{B}_{\sigma(t)} \mathcal{K}_{\sigma(t)} \hat{\mathcal{X}}(t - \tau(t)) + \mathcal{B}_{\sigma(t)} \mathcal{K}_{\sigma(t)} \hat{e}(i_k h) \\
&\quad + \mathcal{B}_{\mathcal{G}\sigma(t)} \mathcal{G}\left(\hat{\mathcal{X}}(t)\right) + \mathcal{L}_{\sigma(t)} \mathcal{C}_{\sigma(t)} \Delta\mathcal{X}(t), \\
\Delta\dot{\mathcal{X}}(t) &= \left(\mathcal{A}_{\sigma(t)} - \mathcal{L}_{\sigma(t)} \mathcal{C}_{\sigma(t)}\right) \Delta\mathcal{X}(t) + \mathcal{B}_{\mathcal{G}\sigma(t)} \mathfrak{G}(t) - \mathcal{B}_{\sigma(t)} \mathcal{C}_{\mathcal{F}\sigma(t)} \mathcal{X}_{\mathcal{F}}(t), \\
\dot{\mathcal{X}}_{\mathcal{F}}(t) &= \left(\mathcal{A}_{\mathcal{F}\sigma(t)} + \mathcal{B}_{\mathcal{F}\sigma(t)} \mathcal{C}_{\mathcal{F}\sigma(t)}\right) \mathcal{X}_{\mathcal{F}}(t) + \mathcal{B}_{\mathcal{F}\sigma(t)} \mathcal{B}^{+}_{\sigma(t)} \mathcal{L}_{\sigma(t)} \mathcal{C}_{\sigma(t)} \Delta\mathcal{X}(t).
\end{aligned} \tag{10}$$

where $\mathfrak{G}(t) = \mathcal{G}(\mathcal{X}(t)) - \mathcal{G}\left(\hat{\mathcal{X}}(t)\right)$ and the state-space representation of the closed-loop system is

$$\dot{\varphi}(t) = \mathcal{A}_{\sigma(t)} \varphi(t) + \mathcal{B}_{1\sigma(t)} \varphi(t - \tau(t)) + \mathcal{B}_{2\sigma(t)} \hat{e}(i_k h) + \mathcal{B}_{3\sigma(t)} \varphi(\mathcal{G}(\mathcal{X}(t))) \tag{11}$$

and the system matrices are given as

$$\mathcal{A}_{\sigma(t)} = \begin{bmatrix} \mathcal{A}_{\sigma(t)} & \mathcal{L}_{\sigma(t)}\mathcal{C}_{\sigma(t)} & 0 \\ 0 & \mathcal{A}_{\sigma(t)} - \mathcal{L}_{\sigma(t)}\mathcal{C}_{\sigma(t)} & -\mathcal{B}_{\sigma(t)}\mathcal{C}_{\mathcal{F}\sigma(t)} \\ 0 & \mathcal{B}_{\mathcal{F}\sigma(t)}\mathcal{B}^{+}_{\sigma(t)}\mathcal{L}_{\sigma(t)}\mathcal{C}_{\sigma(t)} & \mathcal{A}_{\mathcal{F}\sigma(t)} + \mathcal{B}_{\mathcal{F}\sigma(t)}\mathcal{C}_{\mathcal{F}\sigma(t)} \end{bmatrix},$$

$$\mathcal{B}_{1\sigma(t)} = \begin{bmatrix} \mathcal{B}_{\sigma(t)}\mathcal{K}_{\sigma(t)} & 0 & 0 \\ 0 & 0 & 0 \\ 0 & 0 & 0 \end{bmatrix}, \mathcal{B}_{2\sigma(t)} = \begin{bmatrix} \mathcal{B}_{\sigma(t)}\mathcal{K}_{\sigma(t)} \\ 0 \\ 0 \end{bmatrix},$$

$$\mathcal{B}_{3\sigma(t)} = \begin{bmatrix} \mathcal{B}_{\mathcal{G}\sigma(t)} & 0 & 0 \\ 0 & \mathcal{B}_{\mathcal{G}\sigma(t)} & 0 \\ 0 & 0 & 0 \end{bmatrix}, \varphi(\mathcal{G}(\mathcal{X}(t))) = \begin{bmatrix} \mathcal{G}(\mathcal{X}(t)) \\ \mathfrak{G}(t) \\ 0 \end{bmatrix}.$$

Moreover, the upcoming definitions and lemma are more significant for substantiating the required results in the forthcoming section.

Definition 2 [7] For any $\mathcal{T}_2 > \mathcal{T}_1 > 0$, let $\mathcal{N}_\sigma(\mathcal{T}_1, \mathcal{T}_2)$ denote the switching number of $\sigma(t)$ on an interval $(\mathcal{T}_1, \mathcal{T}_2)$. If $\mathcal{N}_\sigma(\mathcal{T}_1, \mathcal{T}_2) \leq N_0 + \frac{\mathcal{T}_2 - \mathcal{T}_1}{\tau_a}$, holds for given $\tau_a > 0$ and $N_0 \geq 0$, then τ_a is the ADT, and $\mathcal{N}_0$ be the chatter bound.

Definition 3 [7] The augmented error system Eq. (11) is said to be exponentially stable under the switching signal $\sigma(t)$, if there exist two constants $\mathcal{M} > 0$ and $\aleph > 0$ such that the following inequality holds:

$$\|\varphi(t)\| \leq \mathcal{M}e^{-\aleph(t-t_0)}\|\phi\|_L, \quad t \geq t_0, \tag{12}$$

$\|\phi\|_L = sup_{0 \leq \theta \leq t_0}\|\varphi(\theta)\|$ and $\aleph$ is the decay rate.

Lemma 1 [25] Assume $\tau(t) \in [0, \tau_M]$, for any matrices $\mathcal{R}_i \in \mathbb{R}^{n\times n}$ and $\mathcal{M}_i \in \mathbb{R}^{n\times n}$ that satisfy $\begin{bmatrix} \mathcal{R}_i & \mathcal{M}_i \\ \mathcal{M}_i^T & \mathcal{R}_i \end{bmatrix} \geq 0$, the following inequality holds:

$$-\tau_M \int_{t-\tau_M}^{t} \dot{x}^T(s)\mathcal{R}_i\dot{x}(s)ds \leq \zeta_1^T(t)\Gamma_i\zeta(t) \tag{13}$$

where

$$\zeta_1(t) = \begin{bmatrix} x(t) \\ x(t-\tau(t)) \\ x(t-\tau_M) \end{bmatrix}, \Gamma_i = \begin{bmatrix} -\mathcal{R}_i & * & * \\ \mathcal{R}_i^T - \mathcal{M}_i^T & -2\mathcal{R}_i + \mathcal{M}_i + \mathcal{M}_i^T & * \\ \mathcal{M}_i^T & \mathcal{R}_i^T - \mathcal{M}_i^T & -\mathcal{R}_i \end{bmatrix}.$$

3. Main results

The prime intention of this section is to design the observer-based event-triggered scheme for SNNCSs by dint of EID technique. More precisely, based on the appropriate Lyapunov stability theory, we establish a new set of sufficient criterion for the existence of observer based disturbance rejection event-triggered control designs that can be expressed in terms of LMIs, which makes the augmented error system Eq. (11)

is exponential stable. For analytical convenience, we define the positive definite matrices as follows: $\mathcal{P}_i = \text{diag}\{\mathcal{P}_{1i}, \mathcal{P}_{2i}, \mathcal{P}_{3i}\}$, $\mathcal{Q}_i = \text{diag}\{\mathcal{Q}_{1i}, \mathcal{Q}_{2i}, \mathcal{Q}_{3i}\}$, $\mathcal{R}_i = \text{diag}\{\mathcal{R}_{1i}, \mathcal{R}_{2i}, \mathcal{R}_{3i}\}$.

3.1 Average dwell-time analysis

In this subsection, we derive sufficient conditions for the exponential stability of the closed-loop SNNCS Eq. (11) along with observer-based disturbance rejection event-triggered scheme by using the average dwell time technique.

Theorem 3.1 For given trigger parameter ς_i, nonnegative real scalar τ_M, and parameters α, μ, the controller gain matrices $\mathcal{K}_i, \mathcal{L}_i$, the closed-loop SNNCS Eq. (11) with event-triggered sampling scheme is exponentially stable if there exist matrices $\mathcal{P}_i, \mathcal{Q}_i, \mathcal{R}_i, \Psi_{1i}, \Psi_{2i}$ and scalars $\beta_{\mathcal{G}}, \beta_{\mathcal{H}}$ such that the following LMIs hold for all $i \in \mathcal{I}$:

$$\Theta_i = \begin{bmatrix} \Theta_{i,1,1} & \Theta_{i,1,2} & \cdots & \Theta_{i,1,17} \\ \Theta_{i,2,1} & \Theta_{i,2,2} & \cdots & \Theta_{i,2,17} \\ \vdots & \vdots & & \vdots \\ \Theta_{i,17,1} & \Theta_{i,17,2} & \cdots & \Theta_{i,17,17} \end{bmatrix} < 0, \tag{14}$$

where,

$$\begin{aligned}
&\Theta_{i,1,1} = \alpha\mathcal{P}_{1i} + \mathcal{Q}_{1i} - e^{-\alpha\tau_M}\mathcal{R}_{1i}, \Theta_{i,2,2} = \alpha\mathcal{P}_{2i} + \mathcal{Q}_{2i} - e^{-\alpha\tau_M}\mathcal{R}_{2i},\\
&\Theta_{i,3,3} = \alpha\mathcal{P}_{3i} + \mathcal{Q}_{3i} - e^{-\alpha\tau_M}\mathcal{R}_{3i}, \Theta_{i,4,1} = e^{-\alpha\tau_M}(\mathcal{R}_{1i} - \mathcal{M}_{1i}),\\
&\Theta_{i,4,4} = \varsigma_i\Psi_{2i} + e^{-\alpha\tau_M}\left(-2\mathcal{R}_{1i} + \mathcal{M}_{1i} + +\mathcal{M}_{1i}^T\right), \Theta_{i,5,2} = e^{-\alpha\tau_M}(\mathcal{R}_{2i} - \mathcal{M}_{2i}),\\
&\Theta_{i,5,5} = e^{-\alpha\tau_M}\left(-2\mathcal{R}_{2i} + \mathcal{M}_{2i} + \mathcal{M}_{2i}^T\right), \Theta_{i,6,3} = e^{-\alpha\tau_M}(\mathcal{R}_{3i} - \mathcal{M}_{3i}),\\
&\Theta_{i,6,6} = e^{-\alpha\tau_M}\left(-2\mathcal{R}_{3i} + \mathcal{M}_{3i} + \mathcal{M}_{3i}^T\right), \Theta_{i,7,1} = e^{-\alpha\tau_M}\mathcal{M}_{1i},\\
&\Theta_{i,7,4} = e^{-\alpha\tau_M}(\mathcal{R}_{1i} - \mathcal{M}_{1i}), \Theta_{i,7,7} = -e^{-\alpha\tau_M}\mathcal{Q}_{1i} - e^{-\alpha\tau_M}\mathcal{R}_{1i},\\
&\Theta_{i,8,2} = e^{-\alpha\tau_M}\mathcal{M}_{2i}, \Theta_{i,8,5} = e^{-\alpha\tau_M}(\mathcal{R}_{2i} - \mathcal{M}_{2i}), \Theta_{i,8,8} = -e^{-\alpha\tau_M}\mathcal{Q}_{2i} - e^{-\alpha\tau_M}\mathcal{R}_{2i},\\
&\Theta_{i,9,3} = e^{-\alpha\tau_M}\mathcal{M}_{3i}, \Theta_{i,9,6} = e^{-\alpha\tau_M}(\mathcal{R}_{3i} - \mathcal{M}_{3i}), \Theta_{i,9,9} = -e^{-\alpha\tau_M}\mathcal{Q}_{3i} - e^{-\alpha\tau_M}\mathcal{R}_{3i},\\
&\Theta_{i,10,1} = 2\mathcal{P}_{1i}^T + 2\mathcal{N}_{1i}\mathcal{A}_i, \Theta_{i,10,2} = 2\mathcal{N}_{1i}\mathcal{L}_i\mathcal{C}_i, \Theta_{i,10,4} = 2\mathcal{N}_{1i}\mathcal{B}_i\mathcal{K}_i,\\
&\Theta_{i,10,10} = \tau_M^2\mathcal{R}_{1i} - 2\mathcal{N}_{1i}, \Theta_{i,11,2} = 2\mathcal{P}_{2i}^T + 2\mathcal{N}_{2i}\mathcal{A}_i - 2\mathcal{N}_{2i}\mathcal{L}_i\mathcal{C}_i,\\
&\Theta_{i,11,3} = -2\mathcal{N}_{2i}\mathcal{B}_i\mathcal{C}_{\mathcal{F}i}, \Theta_{i,11,11} = \tau_M^2\mathcal{R}_{2i} - 2\mathcal{N}_{2i}, \Theta_{i,12,2} = 2\mathcal{N}_{3i}\mathcal{B}_{\mathcal{F}i}\mathcal{B}_i^+\mathcal{L}_i\mathcal{C}_i,\\
&\Theta_{i,12,3} = 2\mathcal{P}_{3i}^T + 2\mathcal{N}_{3i}\mathcal{A}_{\mathcal{F}i} + 2\mathcal{N}_{3i}\mathcal{B}_{\mathcal{F}i}\mathcal{C}_{\mathcal{F}i}, \Theta_{i,12,12} = \tau_M^2\mathcal{R}_{3i} - 2\mathcal{N}_{3i},\\
&\Theta_{i,13,10} = 2\mathcal{B}_{\mathcal{F}i}^T\mathcal{N}_{1i}^T, \Theta_{i,13,13} = -\beta_{\mathcal{G}}, \Theta_{i,14,11} = 2\mathcal{B}_{\mathcal{F}i}^T\mathcal{N}_{2i}^T, \Theta_{i,14,14} = -\beta_{\mathcal{H}},\\
&\Theta_{i,15,10} = 2\mathcal{K}_i^T\mathcal{B}_i^T\mathcal{N}_{1i}^T, \Theta_{i,15,15} = -\Psi_{1i}, \Theta_{i,16,1} = \mathcal{G}^T, \Theta_{i,16,16} = -\beta_{\mathcal{G}},\\
&\Theta_{i,17,2} = \mathcal{G}^T, \Theta_{i,17,17} = -\beta_{\mathcal{H}},
\end{aligned}$$

and the remaining terms are zero. Then the average dwell-time scheme

$$\tau_a > \tau_a^* = \frac{\ln\mu}{\alpha}, \tag{15}$$

where $\mu \geq 1$ satisfies (for all $i, j \in \mathcal{I}$)

$$\mathcal{P}_i \leq \mu\mathcal{P}_j, \mathcal{Q}_i \leq \mu\mathcal{Q}_j, \mathcal{R}_i \leq \mu\mathcal{R}_j. \tag{16}$$

Moreover, the estimation of state decay is given by

$$\|\varphi(t)\| \leq \sqrt{\frac{\beta_2}{\beta_1}} e^{-\aleph(t-t_0)} \|\varphi(t_0)\|_L, \quad \sqrt{\frac{\beta_2}{\beta_1}} \geq 1 \tag{17}$$

where

$$\aleph = \frac{1}{2}\left(\alpha - \frac{\ln \mu}{\tau_a}\right), \beta_1 = \min_{i \in \mathcal{I}} \lambda_{\max}(\mathcal{P}_i),$$

$$\beta_2 = \max_{i \in \mathcal{I}} \lambda_{\max}(\mathcal{P}_i) + \tau_M e^{-\alpha\tau_M} \max_{i \in \mathcal{I}} \lambda_{\max}(\mathcal{Q}_i) + (\tau_M^2/2) e^{-\alpha\tau_M} \max_{i \in \mathcal{I}} \lambda_{\max}(\mathcal{R}_i).$$

Proof: To achieve the desired result, the LKF for the closed-loop SNNCS Eq. (11) is constructed in the following form:

$$\begin{aligned} \mathcal{V}_i(\varphi(t)) = \varphi^T(t)\mathcal{P}_i\varphi(t) + \int_{t-\tau_M}^{t} e^{\alpha(s-t)} \varphi^T(s)\mathcal{Q}_i\varphi(s)ds \\ +\tau_M \int_{t-\tau_M}^{t} \int_{\theta}^{t} e^{\alpha(s-t)} \dot{\varphi}^T(s)\mathcal{R}_i\dot{\varphi}(s)dsd\theta. \end{aligned} \tag{18}$$

Then, computing the time derivatives of $V_i(\varphi(t))$ along the trajectories of SNNCS Eq. (11), we can get

$$\begin{aligned} \dot{\mathcal{V}}_i(\varphi(t)) + \alpha\mathcal{V}_i(\varphi(t)) = {} & 2\varphi^T(t)\mathcal{P}_i\dot{\varphi}(t) + \varphi^T(t)(\alpha\mathcal{P}_i + \mathcal{Q}_i)\varphi(t) \\ & -e^{-\alpha\tau_M}\varphi^T(t-\tau_M)\mathcal{Q}_i\varphi(t-\tau_M) + \tau_M^2\dot{\varphi}^T(t)\mathcal{R}_i\dot{\varphi}(t) \\ & -\tau_M e^{-\alpha\tau_M} \int_{t-\tau_M}^{t} \dot{\varphi}^T(s)\mathcal{R}_i\dot{\varphi}(s)ds. \end{aligned} \tag{19}$$

According to Lemma 1 for matrices $\mathcal{M}_{ia}(a = 1,2,3)$, the integral terms in Eq. (19) can be expressed as

$$-\tau_M e^{-\alpha\tau_M} \int_{t-\tau_M}^{t} \dot{\varphi}^T(s)\mathcal{R}_i\dot{\varphi}(s)ds \leq e^{-\alpha\tau_M} \zeta_a^T(t)\Gamma_a\zeta_a(t), \quad (a = 1,2,3) \tag{20}$$

where

$$\zeta_1(t) = \begin{bmatrix} x(t) \\ x(t-\tau(t)) \\ x(t-\tau_M) \end{bmatrix}, \zeta_2(t) = \begin{bmatrix} \Delta(x(t)) \\ \Delta(x(t-\tau(t))) \\ \Delta(x(t-\tau_M)) \end{bmatrix}, \zeta_3(t) = \begin{bmatrix} \mathcal{X}_{\mathcal{F}}(t) \\ \mathcal{X}_{\mathcal{F}}(t-\tau(t)) \\ \mathcal{X}_{\mathcal{F}}(t-\tau_M) \end{bmatrix},$$

$$\Gamma_a = \begin{bmatrix} -\mathcal{R}_{ia} & * & * \\ \mathcal{R}_{ia} - \mathcal{M}_{ia} & -2\mathcal{R}_{ia} + \mathcal{M}_{ia} + \mathcal{M}_{ia}^T & * \\ \mathcal{M}_{ia} & \mathcal{R}_{ia} - \mathcal{M}_{ia} & -\mathcal{R}_{ia} \end{bmatrix}.$$

Furthermore, it is to be specified that

$$\beta_{\mathcal{G}}^{-1}\mathcal{X}^T(t)\mathcal{G}^T\mathcal{G}\mathcal{X}(t) - \beta_{\mathcal{G}}\mathcal{G}^T(\mathcal{X}(t))\mathcal{G}(\mathcal{X}(t)) > 0, \tag{21}$$

$$\beta_{\mathcal{H}}^{-1}\Delta^T(\mathcal{X}(t))\mathcal{G}^T\mathcal{G}\Delta(\mathcal{X}(t))-\beta_{\mathcal{H}}\mathfrak{G}^T(t)\mathfrak{G}(t)>0. \tag{22}$$

On the other hand, for any matrices $\mathcal{N}_i$ the following equality holds

$$\dot{\varphi}^T(t)\mathcal{N}_i[\mathcal{A}_i\varphi(t)+\mathcal{B}_{1i}\varphi(t-\tau(t))+\mathcal{B}_{2i}\hat{e}(i_kh)+\mathcal{B}_{3i}\varphi(\mathcal{G}(\mathcal{X}(t)))-\dot{\varphi}(t)]=0 \tag{23}$$

where $\mathcal{N}_i=\begin{bmatrix}\mathcal{N}_{1i}^T & \mathcal{N}_{2i}^T & \mathcal{N}_{3i}^T\end{bmatrix}^T$.

By unifying Eq. (18)-Eq. (23) along with Eq. (7) and applying Schur complement, it can be easy to obtain that

$$\begin{aligned}\dot{\mathcal{V}}_i(\varphi(t))+\alpha\mathcal{V}_i(\varphi(t))-\hat{e}^T(i_kh)\Psi_{1i}\hat{e}(i_kh)+\varsigma_i\hat{x}^T(t-\tau(t))\Psi_{2i}\hat{x}(t-\tau(t))\\ =\Xi^T(t)\Theta\Xi(t)\end{aligned} \tag{24}$$

where $\Xi^T(t)=\begin{bmatrix}\varphi^T(t) & \varphi^T(t-\tau(t)) & \varphi^T(t-\tau_M) & \dot{\varphi}^T(t) & \mathcal{G}^T(\mathcal{X}(t)) & \mathfrak{G}^T(t) & \hat{e}^T(i_kh)\end{bmatrix}^T$ and the element of Θ is detailed in Eq. (14).

Similar to the work in [26] the inequality Eq. (24) can equivalently be rephrased in the following form:

$$\dot{\mathcal{V}}_i(\varphi(t))+\alpha\mathcal{V}_i(\varphi(t))\leq 0. \tag{25}$$

Suppose, $t\in[t_k,t_{k+1}]$ and from Eq. (18), we can get

$$\mathcal{V}_i(\varphi(t))\leq e^{-\alpha(t-t_k)}\mathcal{V}_i(\varphi(t_k)). \tag{26}$$

From Eq. (16), at switching instant t_k, we have $\mathcal{V}_i(\varphi(t_i))\leq\mu\mathcal{V}_{i(t_k^-)}(\varphi(t_k^-))$. Then, it follows from Eq. (10) and the relation $\rho=\mathcal{N}_\sigma(t_0,t)\leq\frac{t-t_0}{\tau_a}, t_0=0$ that

$$\begin{aligned}\mathcal{V}_i(\varphi(t_i)) &\leq e^{-\alpha(t-t_k)}\mu\mathcal{V}_{i(t_k^-)}(\varphi(t_k^-))\\ &\leq e^{-\alpha(t-t_k)}\mu^2\mathcal{V}_{i(t_k^-)}(\varphi(t_k^-))\\ &\cdots\\ &\leq e^{-\alpha(t-t_k)}\mu^\rho\mathcal{V}_{i(t_0)}(\varphi(t_0))\\ &\leq e^{-\left(\alpha-\frac{ln\mu}{\tau_a}\right)(t-t_0)}\mathcal{V}_{i(t_0)}(\varphi(t_0)).\end{aligned} \tag{27}$$

Furthermore, from Eq. (18) and Eq. (27), we can get

$$\begin{aligned}&\beta_1\|\varphi(t)\|^2\leq\mathcal{V}_i(\varphi(t))\leq e^{-\left(\alpha-\frac{ln\mu}{\tau_a}\right)(t-t_0)}\mathcal{V}_i(\varphi(t_0),i(t_0))\leq\beta_2\|\varphi(t_0)\|_L^2,\\ &\|\varphi(t)\|^2\leq\frac{\beta_2}{\beta_1}e^{-\left(\alpha-\frac{ln\mu}{\tau_a}\right)(t-t_0)}\|\varphi(t_0)\|_L^2,\\ &\|\varphi(t)\|\leq\sqrt{\frac{\beta_2}{\beta_1}}e^{-\aleph(t-t_0)}\|\varphi(t_0)\|_L.\end{aligned} \tag{28}$$

Hence, by using τ_a one can easily obtain $\aleph<1$. Then, from Definition 3, it can be concluded that the SNNCS Eq. (11) is exponentially stable. The proof is now completed.

Theorem 3.2 For given trigger parameter ς_i, nonnegative real scalar τ_M, parameters α, μ, and a very small $\rho > 0$, the closed-loop SNNCS Eq. (11) with event-triggered sampling scheme is exponentially stabilizable if there exist matrices $\mathcal{X}_i, \tilde{\mathcal{Q}}_i, \tilde{\mathcal{R}}_i, \tilde{\Psi}_{1i}, \tilde{\Psi}_{2i}$ appropriate dimensioned matrices $\mathcal{Y}_i, \mathcal{Z}_i$, and scalars $\beta_{\mathcal{G}}, \beta_{\mathcal{H}}$ such that the following LMIs hold for all $i \in \mathcal{I}$:

$$\tilde{\Theta}_i = \begin{bmatrix} \tilde{\Theta}_{i,1,1} & \tilde{\Theta}_{i,1,2} & \cdots & \tilde{\Theta}_{i,1,17} \\ \tilde{\Theta}_{i,2,1} & \tilde{\Theta}_{i,2,2} & \cdots & \tilde{\Theta}_{i,2,17} \\ \vdots & \vdots & & \vdots \\ \tilde{\Theta}_{i,17,1} & \tilde{\Theta}_{i,17,2} & \cdots & \tilde{\Theta}_{i,17,17} \end{bmatrix} < 0, \tag{29}$$

$$\begin{bmatrix} -\rho I & C_i^T \overline{\mathcal{X}}_{2i} - \mathcal{X}_{2i} C_i^T \\ 0 & -I \end{bmatrix} < 0 \tag{30}$$

where

$$\begin{aligned}
&\tilde{\Theta}_{i,1,1} = \alpha \mathcal{X}_{1i} + \tilde{\mathcal{Q}}_{1i} - e^{-\alpha\tau_M} \tilde{\mathcal{R}}_{1i}, \tilde{\Theta}_{i,2,2} = \alpha \mathcal{P}_{2i} + \tilde{\mathcal{Q}}_{2i} - e^{-\alpha\tau_M} \tilde{\mathcal{R}}_{2i},\\
&\tilde{\Theta}_{i,3,3} = \alpha \mathcal{P}_{3i} + \tilde{\mathcal{Q}}_{3i} - e^{-\alpha\tau_M} \tilde{\mathcal{R}}_{3i}, \tilde{\Theta}_{i,4,1} = e^{-\alpha\tau_M} \left(\tilde{\mathcal{R}}_{1i} - \tilde{\mathcal{M}}_{1i}\right),\\
&\tilde{\Theta}_{i,4,4} = \varsigma_i \tilde{\Psi}_{2i} + e^{-\alpha\tau_M} \left(-2\tilde{\mathcal{R}}_{1i} + \tilde{\mathcal{M}}_{1i} + \tilde{\mathcal{M}}_{1i}^T\right), \tilde{\Theta}_{i,5,2} = e^{-\alpha\tau_M} \left(\tilde{\mathcal{R}}_{2i} - \tilde{\mathcal{M}}_{2i}\right),\\
&\tilde{\Theta}_{i,5,5} = e^{-\alpha\tau_M} \left(-2\tilde{\mathcal{R}}_{2i} + \tilde{\mathcal{M}}_{2i} + \tilde{\mathcal{M}}_{2i}^T\right), \tilde{\Theta}_{i,6,3} = e^{-\alpha\tau_M} \left(\tilde{\mathcal{R}}_{3i} - \tilde{\mathcal{M}}_{3i}\right),\\
&\tilde{\Theta}_{i,6,6} = e^{-\alpha\tau_M} \left(-2\tilde{\mathcal{R}}_{3i} + \tilde{\mathcal{M}}_{3i} + \tilde{\mathcal{M}}_{3i}^T\right), \tilde{\Theta}_{i,7,1} = e^{-\alpha\tau_M} \tilde{\mathcal{M}}_{1i},\\
&\tilde{\Theta}_{i,7,4} = e^{-\alpha\tau_M} \left(\tilde{\mathcal{R}}_{1i} - \tilde{\mathcal{M}}_{1i}\right), \tilde{\Theta}_{i,7,7} = -e^{-\alpha\tau_M} \tilde{\mathcal{Q}}_{1i} - e^{-\alpha\tau_M} \tilde{\mathcal{R}}_{1i},\\
&\tilde{\Theta}_{i,8,2} = e^{-\alpha\tau_M} \tilde{\mathcal{M}}_{2i}, \tilde{\Theta}_{i,8,5} = e^{-\alpha\tau_M} \left(\tilde{\mathcal{R}}_{2i} - \tilde{\mathcal{M}}_{2i}\right), \tilde{\Theta}_{i,8,8} = -e^{-\alpha\tau_M} \tilde{\mathcal{Q}}_{2i} - e^{-\alpha\tau_M} \tilde{\mathcal{R}}_{2i},\\
&\tilde{\Theta}_{i,9,3} = e^{-\alpha\tau_M} \tilde{\mathcal{M}}_{3i}, \tilde{\Theta}_{i,9,6} = e^{-\alpha\tau_M} \left(\tilde{\mathcal{R}}_{3i} - \tilde{\mathcal{M}}_{3i}\right), \tilde{\Theta}_{i,9,9} = -e^{-\alpha\tau_M} \tilde{\mathcal{Q}}_{3i} - e^{-\alpha\tau_M} \tilde{\mathcal{R}}_{3i},\\
&\tilde{\Theta}_{i,10,1} = 2\mathcal{X}_{1i}^T + 2\beta_{1i} \mathcal{A}_i \mathcal{X}_i, \tilde{\Theta}_{i,10,2} = 2\beta_{1i} \mathcal{C}_i \mathcal{Z}_i, \tilde{\Theta}_{i,10,4} = 2\beta_{1i} \mathcal{B}_i \mathcal{Y}_i,\\
&\tilde{\Theta}_{i,10,10} = \tau_M^2 \tilde{\mathcal{R}}_{1i} - 2\beta_{1i} \mathcal{X}_{1i}, \tilde{\Theta}_{i,11,2} = 2\mathcal{X}_{2i}^T + 2\beta_{2i} \mathcal{A}_i \mathcal{X}_{2i} - 2\beta_{2i} \mathcal{C}_i \mathcal{Z}_i,\\
&\tilde{\Theta}_{i,11,3} = -2\beta_{2i} \mathcal{B}_i \mathcal{C}_{\mathcal{F}i} \mathcal{X}_{3i}, \tilde{\Theta}_{i,11,11} = \tau_M^2 \tilde{\mathcal{R}}_{2i} - 2\beta_{2i} \mathcal{X}_{2i},\\
&\tilde{\Theta}_{i,12,2} = 2\beta_{3i} \mathcal{B}_{\mathcal{F}i} \mathcal{B}_i^{+} \mathcal{C}_i \mathcal{Z}_i, \tilde{\Theta}_{i,12,3} = 2\mathcal{X}_{3i}^T + 2\beta_{3i} \mathcal{A}_{\mathcal{F}i} \mathcal{X}_{3i} + 2\beta_{3i} \mathcal{B}_{\mathcal{F}i} \mathcal{C}_{\mathcal{F}i} \mathcal{X}_{3i},\\
&\tilde{\Theta}_{i,12,12} = \tau_M^2 \tilde{\mathcal{R}}_{3i} - 2\beta_{3i} \mathcal{X}_{3i}, \tilde{\Theta}_{i,13,10} = 2\beta_{1i} \mathcal{B}_{\mathcal{F}i}^T, \tilde{\Theta}_{i,13,13} = -\beta_{\mathcal{G}}, \tilde{\Theta}_{i,14,11} = 2\beta_{2i} \mathcal{B}_{\mathcal{F}i}^T,\\
&\tilde{\Theta}_{i,14,14} = -\beta_{\mathcal{H}}, \tilde{\Theta}_{i,15,10} = 2\beta_{1i} \mathcal{Y}_i^T \mathcal{B}_i^T, \tilde{\Theta}_{i,15,15} = -\tilde{\Psi}_{1i}, \tilde{\Theta}_{i,16,1} = \mathcal{G}^T \mathcal{X}_{1i},\\
&\tilde{\Theta}_{i,16,16} = -\beta_{\mathcal{G}}, \tilde{\Theta}_{i,17,2} = \mathcal{G}^T \mathcal{X}_{2i}, \tilde{\Theta}_{i,17,17} = -\beta_{\mathcal{H}},
\end{aligned}$$

and the remaining terms are zero. If the above LMIs are feasible, the state and observer controller gain matrices are computed by $\mathcal{K}_i = \mathcal{Y}_i \mathcal{X}_{1i}^{-1}, \mathcal{L}_i = \mathcal{Z}_i \overline{\mathcal{X}}_{2i}^{-1}$. Then the average dwell-time scheme

$$\tau_a > \tau_a^* = \frac{\ln\ \mu}{\alpha}, \tag{31}$$

where $\mu \geq 1$ satisfies (for all $i, j \in \mathcal{I}$)

$$\mathcal{X}_i \leq \mu \mathcal{X}_j,\ \tilde{\mathcal{Q}}_i \leq \mu \tilde{\mathcal{Q}}_j,\ \tilde{\mathcal{R}}_i \leq \mu \tilde{\mathcal{R}}_j. \tag{32}$$

Moreover, the estimation of state decay is given by

$$\|\varphi(t)\| \leq \sqrt{\frac{\beta_2}{\beta_1}} e^{-\aleph(t-t_0)} \|\varphi(t_0)\|_L,\quad \sqrt{\frac{\beta_2}{\beta_1}} \geq 1 \tag{33}$$

where

$$\aleph = \frac{1}{2}\left(\alpha - \frac{\ln\ \mu}{\tau_a}\right), \beta_1 = \min_{i \in \mathcal{I}} \lambda_{\max}\left(\tilde{\mathcal{X}}_i\right),$$

$$\beta_2 = \max_{i \in \mathcal{I}} \lambda_{\max}\left(\tilde{\mathcal{X}}_i\right) + \tau_M e^{-\alpha \tau_M} \max_{i \in \mathcal{I}} \lambda_{\max}\left(\tilde{\mathcal{Q}}_i\right) + \left(\tau_M^2/2\right) e^{-\alpha \tau_M} \max_{i \in \mathcal{I}} \lambda_{\max}\left(\tilde{\mathcal{R}}_i\right).$$

Proof: The proof of this theorem is obtained by following the similar technique together with the same Lyapunov–Krasovskii functional Eq. (18) as in Theorem 3.1. For obtaining the controller gain matrices, let us define $\mathcal{N}_i = \beta_{ai}\mathcal{P}_{ai}, (a = 1,2,3),$ here β_{ai} are the designing parameter. Then pre- and post- multiplying the matrix Θ_i in Eq. (14) by $\{\mathcal{X}_i, \mathcal{X}_i, \mathcal{X}_i, \mathcal{X}_i, I, I, \mathcal{X}_{1i}, I, I\}$, and its transpose, respectively, where $\mathcal{X}_i = \text{diag}\{\mathcal{X}_{1i}, \mathcal{X}_{2i}, \mathcal{X}_{3i}\}$. Note that if (30) holds for a very small $\rho > 0$, then $\mathcal{C}_i\mathcal{X}_{2i}$ almost equals to $\overline{\mathcal{X}}_{2i}\mathcal{C}_i$. Further, by setting $\mathcal{C}_i\mathcal{X}_{2i} = \overline{\mathcal{X}}_{2i}\mathcal{C}_i, \mathcal{X}_i = \mathcal{P}_i^{-1}, \mathcal{X}_i\mathcal{Q}_i\mathcal{X}_i = \tilde{\mathcal{Q}}_i, \mathcal{X}_i\mathcal{R}_i\mathcal{X}_i = \tilde{\mathcal{R}}_i, \mathcal{X}_i\mathcal{M}_{ai}\mathcal{X}_i = \tilde{\mathcal{M}}_{ai}, (a = 1,2,3), \mathcal{X}_{1i}\Psi_{1i}\mathcal{X}_{1i} = \tilde{\Psi}_{1i}, \mathcal{X}_{1i}\Psi_{2i}\mathcal{X}_{1i} = \tilde{\Psi}_{2i}, \mathcal{Y}_i = \mathcal{K}_i\mathcal{X}_{1i}, \mathcal{Z}_i = \mathcal{L}_i\overline{\mathcal{X}}_{2i}$. The LMI in Eq. (29) can thus be easily obtained. As a consequence, if the LMI-based condition Eq. (29) holds, the closed-loop augmented system Eq. (11) is exponentially stable. This concludes the proof.

Remark 1 It should be noted that the condition $\mathcal{C}_i\mathcal{X}_{2i} = \overline{\mathcal{X}}_{2i}\mathcal{C}_i$ is not a strict LMI and is difficult to deal with using an existing LMI package, such as Matlab. In order to tackle such complication, let us consider the optimization strategy procedure for $\mathcal{C}_i\mathcal{X}_{2i} = \overline{\mathcal{X}}_{2i}\mathcal{C}_i$, it may be appropriately rewritten as $\left(\mathcal{C}_i\mathcal{X}_{2i} - \overline{\mathcal{X}}_{2i}\mathcal{C}_i\right)^T\left(\mathcal{C}_i\mathcal{X}_{2i} - \overline{\mathcal{X}}_{2i}\mathcal{C}_i\right) - \rho^2 I < 0,$ where $\rho > 0$ is very small [27]. Hence, the aforementioned optimization issue can be rewritten as Eq. (30), using the Schur Complement.

4. Simulation results

Example 1 The mass-spring-damping system is one of the most common simplified models that are used in mechanical engineering, such as the human exoskeleton back frame model [28] or the bridge dynamics model [29], etc. This is one of the most practical systems in regard to our lives and technology. Therefore, it is extremely beneficial to study the system of mass-spring-damping in more detail.

In this study, we consider the following mass-spring-damping system Eq. (34) as shown in **Figure 4a**.

$$\mathcal{M}\ddot{x}(t) + \mathcal{R}_1 + \mathcal{R}_2 = \mathcal{U}(t) \tag{34}$$

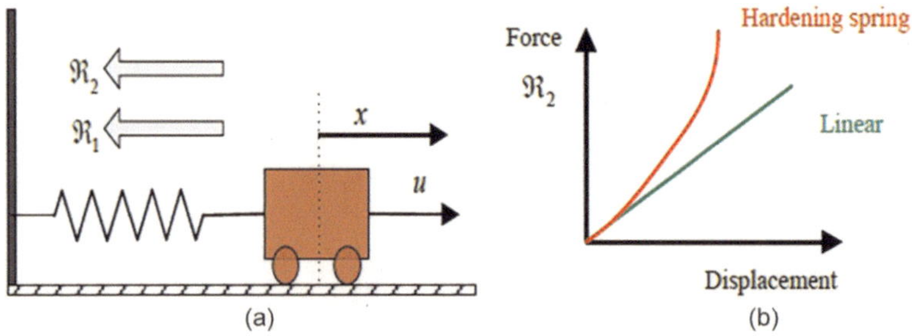

Figure 4.
(a) Mass-spring-damping system, (b) Restoring force $\mathcal{R}_1$ [30].

where $\mathcal{M}$ indicates the mass; $\mathcal{R}_1 = c\dot{x}$ with $c > 0$ and $\mathcal{R}_2$ denotes the friction force and the restoring force, respectively; $\mathcal{U}(t)$ represents for the external input. According to **Figure 4b**, the restoring force $\mathcal{R}_2$ contains a linear component and a hardening spring force. In other words, $\mathcal{R}_2 = kx + ka^2x^3$ with constants k and a, where x signifies the displacement from a reference point.

Further, let $x(t) = [x_1(t) \quad x_2(t)]^T$, in the meantime, consider $a \neq 0$ and $a = 0$ the nonlinear system Eq. (34) could be described by the following two subsystems as given in [30]:

$$\mathcal{A}_1 = \begin{bmatrix} 0 & 1 \\ \dfrac{-k - 4ka^2}{m} & -\dfrac{c}{m} \end{bmatrix}, \mathcal{B}_1 = \begin{bmatrix} 0 \\ \dfrac{1}{m} \end{bmatrix}, \mathcal{A}_2 = \begin{bmatrix} 0 & 1 \\ -\dfrac{k}{m} & -\dfrac{c}{m} \end{bmatrix}, \mathcal{B}_2 = \begin{bmatrix} 0 \\ \dfrac{1}{m} \end{bmatrix}.$$

The following parameter values are employed in this scenario: $m = 1\text{kg}, c = 2\text{N.m/s}, k = 8\text{N/s}$ and $a = 0.3m^{-1}$ [30]. Furthermore, the following system parameters have been listed:

$$\mathcal{B}_{\mathcal{G}1} = \begin{bmatrix} 0.2974 & 0 \\ 0 & 0.2648 \end{bmatrix}, \mathcal{B}_{\mathcal{W}1} = \begin{bmatrix} 0.0737 \\ 0.0637 \end{bmatrix}, \mathcal{C}_1 = \begin{bmatrix} 0.1536 \\ 0.5438 \end{bmatrix}^T,$$

$$\mathcal{B}_{\mathcal{G}2} = \begin{bmatrix} 0.08750 \\ 00.0691 \end{bmatrix}, \mathcal{B}_{\mathcal{W}2} = \begin{bmatrix} 0.1391 \\ 0.0139 \end{bmatrix}, \mathcal{C}_2 = \begin{bmatrix} 0.4167 \\ 0.2345 \end{bmatrix}^T.$$

Let us consider the remaining parameters in Theorem 3.2 to be $\alpha = 0.1, \mu = 1.1, \varsigma_1 = 0.5, \varsigma_2 = 0.5, \rho = 0.5, \beta_{1i} = 0.2114, \beta_{2i} = 0.3835, \beta_{3i} = 0.1885, \tau_M = 0.1,$ the nonlinear term defined as $\mathcal{G}(\mathcal{X}(t)) = 0.1\, tan\, (\mathcal{X}(t))$. Moreover, the low-pass filter parameters are determined as follows: $w_c = 100, \mathcal{A}_{\mathcal{F}i} = -101, \mathcal{B}_{\mathcal{F}i} = 100, \mathcal{C}_{\mathcal{F}i} = 1$. Furthermore, the network-induced exogenous disturbance is addressed by $\mathcal{W}(t) = 0.4 \sin(\pi t/2) + 0.35 \tanh(t - 37) - 0.35 \tanh(t - 27)$. The LMI constraints derived in Eq. (29) are then solved using the MATLAB LMI toolbox, and the feasibility with the aforementioned parameter values can be determined. The feedback and observer gain matrices are provided below based on the parameters discussed in this article:

$$\mathcal{K}_1 = [3.6688 \quad -0.8039], \mathcal{L}_1 = [-5.7747 \quad 0.0034]^T,$$
$$\mathcal{K}_2 = [1.4785 \quad -0.4722], \mathcal{L}_2 = [-4.2569 \quad -0.0143]^T.$$

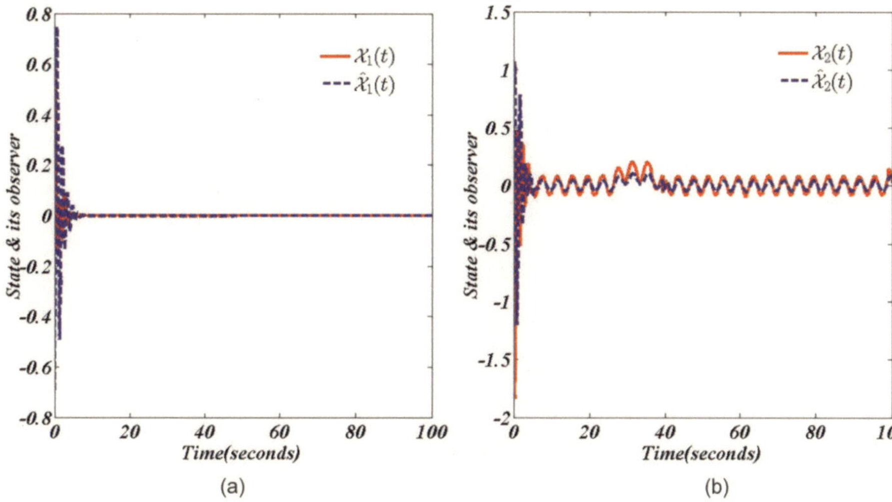

Figure 5.
Simulation results of state and its observer signal. (a) State $X_1(t)$. (b) State $X_2(t)$.

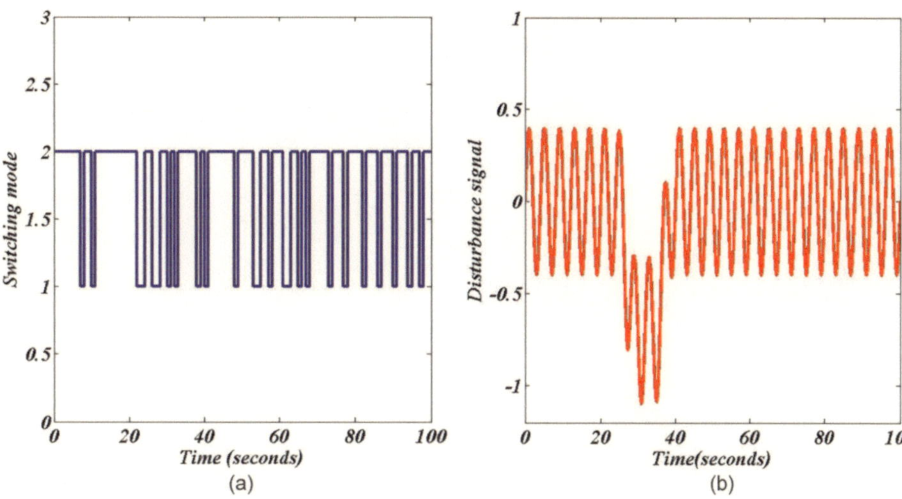

Figure 6.
Simulation results of switching and disturbance signals. (a) Switching signals, (b) disturbance signals.

Furthermore, during the simulation process, the initial values for the state and observer systems are determined as follows: $\mathcal{X}(0) = [0.7 \quad -0.8]^T$ and $\hat{\mathcal{X}}(0) = [-0.7 \quad 0.8]^T$. The simulation results for the SNNCS Eq. (11) are computed from the aforesaid gain values, and the simulation results are displayed in **Figures 5–10**.

So based on the aforementioned controller gain matrices, the real concentration of the actual state and its observer are shown in **Figure 5a** and **Figure 5b**, respectively, concluding that the estimated state closely resembles the actual state. The corresponding switching signal $\sigma(t)$ and the disturbance signals are then displayed **Figure 6a** and **Figure 6b**, respectively. Additionally, the estimated error between real state and their observer is displayed in **Figure 7**. Moreover, **Figure 8a** and **Figure 8b**

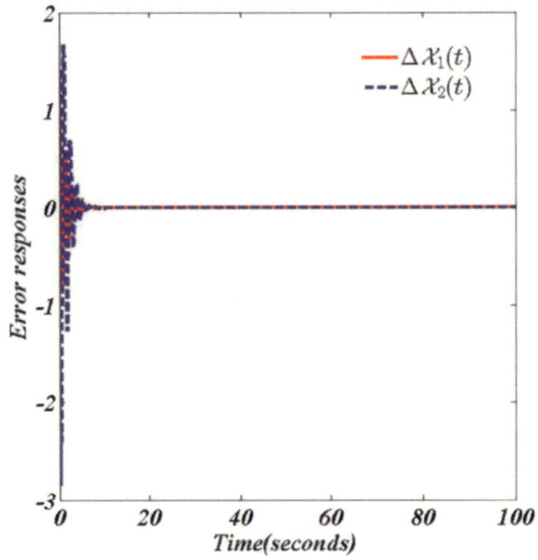

Figure 7.
Simulation results of error signals.

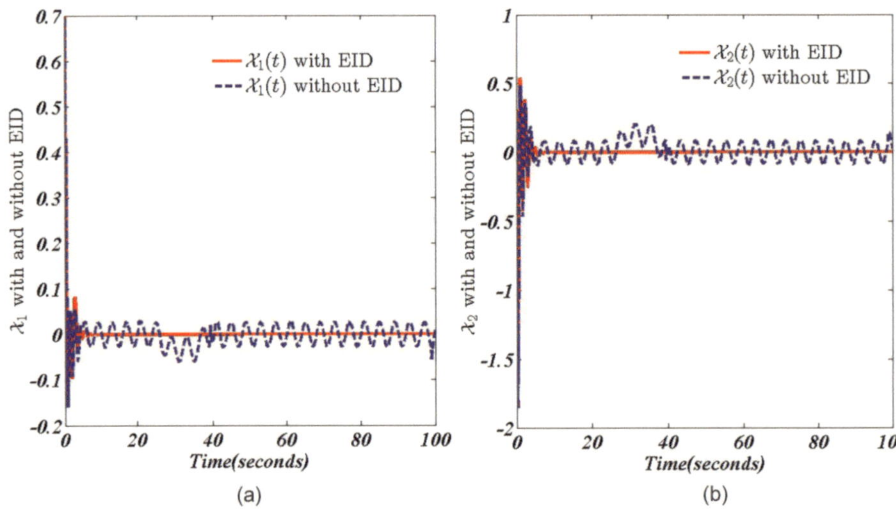

Figure 8.
Simulation results of state responses with and without EID. (a) State $X_1(t)$, (b) State $X_2(t)$.

depict the real state and the observer state in the presence and absence of the EID estimator. The trajectories for an event-triggered release's instants, intervals, error, and threshold values are shown in **Figure 9a** and **Figure 9b**, respectively. The phase portrait of the actual and observer state are then shown in **Figure 10a** and **Figure 10b**, respectively.

Additionally, the ADT is determined to be $\tau_a > \tau_a^* = 0.9531$ for the same parameter values used above. Also, by setting $\tau_a = 1$, the decay rate for the system is determined as $\aleph = 0.0023$ and the state decay is estimated as $\|\varphi(t)\| \leq 1.6003e^{-0.0023(t-t_0)}\|\varphi(t_0)\|_L, \forall t \geq t_0$ which means that the suggested observer-based event-triggered controller ensures the exponential stability of the resulting closed-loop augmented SNNCS Eq. (11).

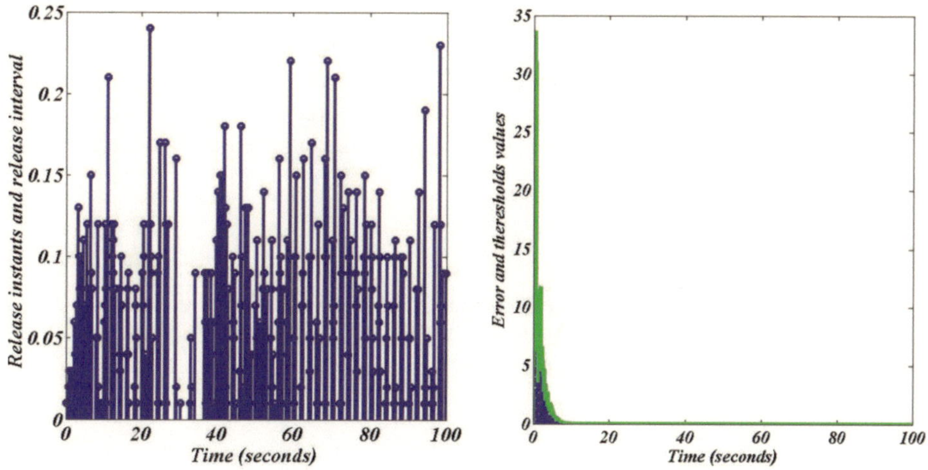

Figure 9.
Responses of release instants and release interval, error, and theresholds for SNNCS Eq. (11).

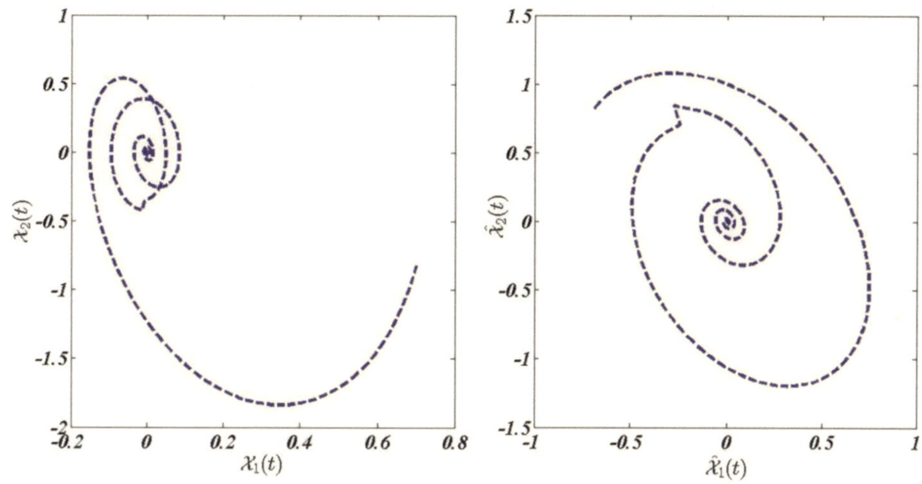

Figure 10.
Phase portrait.

Eventually, we have clearly concluded from the above simulation results that the proposed controller with observer-based event-driven controller has better disturbance rejection performance for the considered SNNCSs.

5. Conclusion

This paper examined an EID-based disturbance rejection approach for SNNCSs with an event-triggered mechanism under observer-based control. A particular focus was placed on the problem of rejecting disturbances in SNNCSs and proposed an EID-based framework to address this issue. By using our method, we are able to reject both matched and unmatched disturbances with no prior knowledge about their characteristics. A new delay-dependent condition is developed utilizing the Lyapunov-Krasovskii functional in conjunction with the ADT technique to ensure the exponential stability of even-triggered closed-loop SNNCSs. It is possible to achieve the appropriate controller gain matrices through the use of LMIs. Ultimately, a numerical example is presented in order to demonstrate the significance of the suggested control strategy.

Acknowledgements

This paper was supported by the National Science and Technology Council of Taiwan under Grant:110-2811-E-019-505-MY2, 110-2221-E-019-073-MY2.

Conflict of interest

The authors declare no conflict of interest.

Author details

Arumugam Arunkumar† and Jenq-Lang Wu*†
Electrical Engineering Department, National Taiwan Ocean University, Keelung, Taiwan, R.O.C

*Address all correspondence to: wujl@mail.ntou.edu.tw

† These authors contributed equally.

References

[1] Fei Z, Guan C, Zhao X. Event-triggered dynamic output feedback control for switched systems with frequent asynchronism. IEEE Transactions on Automatic Control. 2020;**65**(7):3120-3127. DOI: 10.1109/TAC.2019.2945279

[2] Wu M, Gao F, She J, Cao W. Active disturbance rejection in switched neutral-delay systems based on equivalent-input-disturbance approach. IET Control Theory Applications. 2016;**10**:2387-2393. DOI: 10.1049/iet-cta.2016.0211

[3] Chen Z, Zhang Y, Kong Q, Fang T, Wang J. Observer-based control for persistent dwell-time switched networked nonlinear systems under packet dropout. Applied Mathematics and Computation. 2022;**415**(15):126679. DOI: 10.1016/j.amc.2021.126679

[4] Wang Z, Sun J, Chen J. Stability analysis of switched nonlinear systems with multiple time-varying delays. IEEE Transactions on Systems, Man, and Cybernetics: Systems. 2022;**52**(6):3947-3956. DOI: 10.1109/TSMC.2021.3080278

[5] Li T, Fu J, Deng F, Chai T. Stabilization of switched linear neutral systems: An event-triggered sampling control scheme. IEEE Transactions Automatic Control. 2018;**63**:3537-3544. DOI: 10.1109/TAC.2018.2797160

[6] Wu JL. Singular L_2-gain control for switched nonlinear control systems under arbitrary switching. International journal of robust and nonlinear control. 2020;**30**(10):4149-4163. DOI: 10.1002/rnc.4987

[7] Zhao X, Zhang L, Shi P, Liu M. Stability and stabilization of switched linear systems with mode-dependent average dwell time. IEEE Transactions Automatic Control. 2012;**57**(7):1809-1815. DOI: 10.1109/TAC.2011.2178629

[8] Yang SH, Wu J. L: Mixed event/time-triggered static output feedback L_2-gain control for networked control systems. Asian Journal of Control. 2017;**19**(1):1-10. DOI: 10.1002/asjc.1334

[9] Yue D, Tian E, Han QL. A delay system method for designing event-triggered controllers of networked control systems. IEEE Transactions Automatic Control. 2013;**58**(2):475-481. DOI: 10.1109/TAC.2012.2206694

[10] Qi Y, Liu Y, Niu B. Event-triggered H_∞ filtering for networked switched systems with packet disorders. IEEE Transactions on Systems, Man, and Cybernetics: Systems. 2021;**51**(5):2847-2859. DOI: 10.1109/TAC.2012.2206694

[11] Qi Y, Zeng P, Bao W. Event-triggered and self-triggered H_∞ control of uncertain switched linear systems. IEEE Transactions on Systems, Man, and Cybernetics: Systems. 2020;**50**(5):56-64. DOI: 10.1109/TSMC.2018.2801284

[12] Xiao X, Park JH, Zhou L. Event-triggered control of discrete-time switched linear systems with packet losses. Applied Mathematics and Computation. 2018;**333**:344-352. DOI: 10.1016/j.amc.2018.03.122

[13] Jhang JY, Wu JL. Yung C.F: Design of event-triggered state-constrained stabilizing controllers for nonlinear control systems. IEEE Access. 2022;**10**:3659-3667. DOI: 10.1109/ACCESS.2021.3139963

[14] Li L, Fu J, Zhang Y, Chai T, Song L, Albertos P. Output regulation for networked switched systems with alternate event-triggered control under transmission delays and packet losses. Automatica. 2021;**131**:109716. DOI: 10.1016/j.automatica.2021.109716

[15] Gu Z, Yue D, Liu J, Ding Z. H_∞ tracking control of nonlinear networked systems with a novel adaptive event-triggered communication scheme. Journal of the Franklin Institute. 2017; **354**(8):3540-3553. DOI: 10.1016/j.jfranklin.2017.02.020

[16] Zhang D, Xu Z, Karimi HR. Wang Q. G: Distributed filtering for switched linear systems with sensor networks in presence of packet dropouts and quantization. IEEE Transactions on Circuits and Systems I: Regular Papers. 2017;**64**(10):2783-2796. DOI: 10.1109/TCSI.2017.2695481

[17] Gao F, Wu M, She J, He Y. Delay-dependent guaranteed-cost control based on combination of smith predictor and equivalent-input disturbance approach. ISA Transactions. 2016;**62**: 215-221. DOI: 10.1016/j.isatra.2016.02.008

[18] Liu RJ, Liu GP, Wu M, Nie ZY. Disturbance rejection for time-delay systems based on the equivalent-input-disturbance approach. Journal of the Franklin Institute. 2014;**351**:3364-3377. DOI: 10.1016/j.jfranklin.2014.02.015

[19] Gao F, Wu M, She J, Cao W. Disturbance rejection in nonlinear systems based on equivalent-input-disturbance approach. Applied Mathematics and Computation. 2016; **282**:244-253. DOI: 10.1016/j.amc.2016.02.014

[20] Ouyang L, Wu M, She J. Estimation of and compensation for unknown input nonlinearities using equivalent-input-disturbance approach. Nonlinear Dynamics. 2017;**88**:2161-2170. DOI: 10.1007/s11071-017-3369-5

[21] She JH, Fang M, Ohyama Y, Hashimoto H, Wu M. Improving disturbance-rejection performance based on an equivalent-input-disturbance approach. IEEE Transactions on Industrial Electronics. 2008;**55**(1): 380-389. DOI: 10.1109/TIE.2007.905976

[22] Chang XY, Li YL, Zhang WY, Wang N, Xue W. Active disturbance rejection control for a flywheel energy storage system. IEEE Transactions on Industrial Electronics. 2015;**62**(2): 991-1001. DOI: 10.1109/TIE.2014.2336607

[23] Xue W, Bai W, Yang S, Song K, Huang Y, Xie H. ADRC with adaptive extended state observer and its application to air-fuel ratio control in gasoline engines. IEEE Transactions on Industrial Electronics. 2015;**62**(9): 5847-5857. DOI: 10.1109/TIE.2015.2435004

[24] Ren H, Zong G, Li T. Event-triggered finite-time control for networked switched linear systems with asynchronous switching. IEEE Transactions on Systems, Man, and Cybernetics: Systems. 2018;**48**(11): 1874-1884. DOI: 10.1109/TSMC.2017.2789186

[25] Zha L, Tian E, Xie X, Gu Z, Cao J. Decentralized event-triggered H1 control for neural networks subject to cyber-attacks. Information Sciences. 2018;**457-458**:141-155. DOI: 10.1016/j.ins.2018.04.018

[26] Benzaouia A, Eddoukali Y. Robust fault detection and control for continuous-time switched systems with average dwell time. Circuits, Systems,

and Signal Processing. 2018;**37**: 2357-2373. DOI: 10.1007/s00034-017-0674-7

[27] Zhang Y, Liu C. Observer-based finite-time H_∞ control of discrete-time Markovian jump systems. Applied Mathematical Modelling. 2013;**37**: 3748-3760. DOI: 10.1016/j.apm.2012.07.060

[28] Bullimore SR, Burn JF. Ability of the planar spring-mass model to predict mechanical parameters in running humans. Journal of Theoretical Biology. 2007;**248**(4):686-695. DOI: 10.1016/j.jtbi.2007.06.004

[29] Sapountzakis EJ, Syrimi PG, Pantazis IA, Antoniadis IA. KDamper concept in seismic isolation of bridges with flexible piers. Engineering Structures. 2017;**153**:525-539. DOI: 10.1016/j.engstruct.2017.10.044

[30] Li M, Chen Y. Robust adaptive sliding mode control for switched networked control systems with disturbance and faults. IEEE Transactions on Industrial Informatics. 2018;**15**(1):1-11. DOI: 10.1109/TII.2018.2808921

Chapter 5

Robust Control of Space Robots Considering Friction Characteristics

Xinye You

Abstract

In this chapter, a planar two-bar space robot is taken as the research object, and the dynamics equations of fully rigid and joint flexible space robot systems are derived by using momentum conservation and Lagrange equation of the second kind. Aiming at flexible joint space robot system, the system is decomposed into fast and slow subsystems based on singular perturbation. Considering the friction characteristics of joints, a robust controller based on the friction upper bound was designed to offset the influence of friction torque and improve the flat roof phenomenon in joint trajectory tracking control. For flexible joint space robot system, a fast variable subsystem controller is designed using moment differential to actively suppress flexible vibration. The controller of slow subsystem is designed by robust method and a bounded friction compensation term is introduced to offset the influence of joint friction torque. The effectiveness of the proposed manipulator control scheme is verified by numerical simulation experiments.

Keywords: free-floating space robot, flexible joint, robust control, nonlinear friction, friction characteristics

1. Introduction

Space robot systems play an irreplaceable role in the construction of space laboratories, in-orbit maintenance of spacecraft, and recycling of space garbage. Space robots have the ability to perform tasks in the space environment that poses a threat to human life, reducing the risk of debris and strong radiation to astronauts. Therefore, the research on space robots has been widely valued by scholars at home and abroad [1–9].

In low Earth orbit environment, there are various disturbances acting on the space robot system, such as thin atmosphere, joint friction torque, liquid propellant slosh of spacecraft and so on. Therefore, when designing the controller of space robot, it is necessary to consider the influence of system uncertainty and external disturbance, as well as the influence of joint friction torque on the response.

2. Kinematics and dynamics modeling of floating-based space robots

The space robot system is in free-floating state, and its base does not form rigid connection with the inertial system object, so there is a strong coupling effect between the base of the space robot system and each rod of the robot arm.

In order to establish the relationship between trajectory point kinematics and control force, the robot dynamics model should be built. The commonly used multi-body dynamic modeling methods for robot dynamics include Newton-Euler method [1], Kane method [2], Lagrange method of the second kind [3], virtual robotic arm method [4] and Roberson-Weittenberg method [5]. The above modeling method can be extended to rigid space robot system modeling after proper transformation.

In this chapter, the second Lagrange equation and momentum conservation relationship are used to establish a dynamic model of space robot with uncontrolled carrier position and controlled carrier attitude, and the dynamic equation of flexible joint floating space robot system is established based on the simplified linear torsion spring joint model of flexible joint. The work in this chapter lays the foundation for the space robot dynamic control scheme in the following paper.

2.1 Dynamics model of space robot system with attitude controlled but carrier position not controlled

As shown in **Figure 1**, the model is a planar motion fully rigid two-bar space robot consisting of a free-floating carrier $B0$ and two rigid arms $B1$, $B2$.

The symbols involved in formula derivation in **Figure 1** and this section are defined as follows:

OXY: inertial coordinate system;

$O_0X_0Y_0$: space robot carrier coordinate system;

$O_iX_iY_i$: coordinate system of each manipulator $(i = 1, 2)$;

C: total center of mass of space robot system;

Oci: center of mass of each part $Bi(i = 0, 1, 2)$;

$O\mathbf{0}$: the origin of the carrier coordinate system coinciding with the center of mass$Oc0$;

$O\mathbf{1}$, $O2$: respectively are the rotary joint axis of the arm rod $B1$, $B2$;

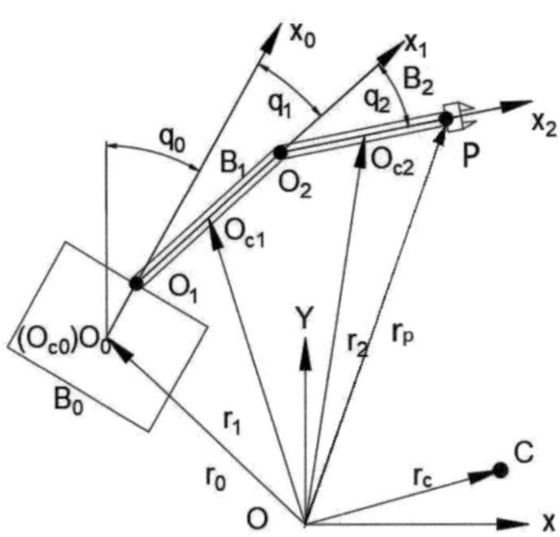

Figure 1.
Free-floating space manipulator system.

x_0: line direction from the origin $O0$ to $O1$ of the carrier coordinate system;
x_i: mechanical arm Bi along the bar direction;
$\mathbf{e}_i$: a unit vector along the axis $x_i (i = 0, 1, 2)$;
$\mathbf{z}_i$: unit vector perpendicular to the axis of each joint of the plane OXY;
$\mathbf{r}_i$: position vector of centroid of each part of inertial system OXY, where $(i = 0, 1, 2)$;
$\mathbf{r}_c$: vector pointing from the origin O of inertial system to the total center of mass C of the system;
l_0: the distance along the axis $O0$ to $O1$;
l_i: length of each arm Bi along the axis $x_i (i = 1, 2)$;
a_1: the distance between the center of mass $Oc1$ of rod 1 and the rotating hinge joint$O1$;
a_2: the distance between the center of mass $Oc2$ of rod 2 and the rotating hinge joint$O2$;
m_i: the mass of each part $Bi(i = 0, 1, 2, 3)$, where m_3 represents the end-load mass;
M: total mass of space robot system, where $M = \sum_{i=0}^{3} m_i$;
Ii: the central inertia tensor of each part $(i = 0, 1, 2, 3)$, where I_3 represents the end-load mass;
Iim: the moment of inertia of the motor at the ith joint;
q_0: attitude Angle of the carrier;
q_i: relative joint Angle of the ith arm $(i = 1, 2)$;
ωi: the rotation angular speed of the carrier and each arm bar $(i = 0, 1, 2)$;

By analyzing the geometric position relation of the system in **Figure 2**, the position vector relation of the center of mass Oci of each part of the floating based rigid space robot in the inertial system OXY can be expressed as follows:

$$\mathbf{r}_1 = \mathbf{r}_0 + l_0\mathbf{e}_0 + a_1\mathbf{e}_1 \tag{1}$$

$$\mathbf{r}_2 = \mathbf{r}_0 + l_0\mathbf{e}_0 + l_1\mathbf{e}_1 + a_2\mathbf{e}_2 \tag{2}$$

where, the position vector $\mathbf{r}_0 = [x_0, y_0]$ of the carrier's centroid, the unit vector $\mathbf{e}_i$along the axis x_i, the attitude Angle of the carrier q_0 and the Angle of each arm joint $q_i (i = 1, 2)$ are expressed as follows:

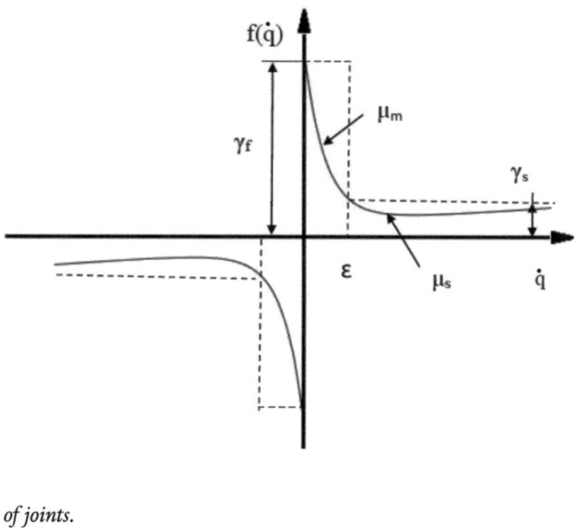

Figure 2.
Friction characteristics of joints.

$$\mathbf{e}_0 = \left[\sin q_0 \quad \cos q_0\right]^{\mathrm{T}} \tag{3}$$

$$\mathbf{e}_1 = \left[\sin(q_0 + q_1) \quad \cos(q_0 + q_1)\right]^{\mathrm{T}} \tag{4}$$

$$\mathbf{e}_2 = \left[\sin(q_0 + q_1 + q_2) \quad \cos(q_0 + q_1 + q_2)\right]^{\mathrm{T}} \tag{5}$$

The velocity vector relationship between the centroid $Oci(i = 0, 1, 2)$ of each component and the terminal P point can be obtained:

$$\dot{\mathbf{r}}_1 = \dot{\mathbf{r}}_0 + l_0\dot{\mathbf{e}}_0 + a_1\dot{\mathbf{e}}_1 \tag{6}$$

$$\dot{\mathbf{r}}_2 = \dot{\mathbf{r}}_0 + l_0\dot{\mathbf{e}}_0 + l_1\dot{\mathbf{e}}_1 + a_2\dot{\mathbf{e}}_2 \tag{7}$$

$$\dot{\mathbf{r}}_p = \dot{\mathbf{r}}_0 + l_0\dot{\mathbf{e}}_0 + l_1\dot{\mathbf{e}}_1 + l_2\dot{\mathbf{e}}_2 \tag{8}$$

$\mathbf{r}_p$: the position vector at the end P of the manipulator in the inertial frame.

Where $\dot{\mathbf{e}}_0 = \dot{q}_0\left[\cos q_0 \quad -\sin q_0\right]^{\mathrm{T}}$, $\dot{\mathbf{e}}_1 = (\dot{q}_0 + \dot{q}_1)\left[\cos(q_0 + q_1) \quad -\sin(q_0 + q_1)\right]^{\mathrm{T}}$, $\dot{\mathbf{e}}_2 = (\dot{q}_0 + \dot{q}_1 + \dot{q}_2)\left[\cos(q_0 + q_1 + q_2) \quad -\sin(q_0 + q_1 + q_2)\right]^{\mathrm{T}}$.

The relationship between the velocity of P point at the end of the deduced arm and the generalized velocity $\dot{\mathbf{q}}_{p0}$ in the inertial system is as follows:

$$\dot{\mathbf{r}}_p = \begin{bmatrix} \dot{x}_p \\ \dot{y}_p \end{bmatrix} = \begin{bmatrix} J_{011} & J_{012} & J_{013} & J_{014} & J_{015} \\ J_{021} & J_{022} & J_{023} & J_{024} & J_{025} \end{bmatrix} \begin{bmatrix} \dot{x}_0 \\ \dot{y}_0 \\ \dot{q}_0 \\ \dot{q}_1 \\ \dot{q}_2 \end{bmatrix} = \mathbf{J}_0\dot{\mathbf{q}}_{p0} \tag{9}$$

The generalized velocity $\dot{\mathbf{q}}_{p0} = \left[\dot{x}_0, \dot{y}_0, \dot{q}_0, \dot{q}_1, \dot{q}_2\right]^{\mathrm{T}}$, $\mathbf{J}_0$ is the Jacobian velocity matrix of the floating space robot system with fully controllable position and attitude.

Considering that the space robot studied in this chapter is based on the assumption that the attitude of the carrier is controlled and its position is not controlled; that is, the attitude of the carrier is controlled by the reaction wheel driven by the motor, and the system itself is not controlled by external forces. The free-floating space robot system follows the momentum conservation relationship without the weak gravity effect. From the definition of the total centroid of the system, we can see the following relation:

$$M\mathbf{r}_c = \sum_{i=0}^{2} m_i\mathbf{r}_i \tag{10}$$

$\mathbf{r}_c = 0$, the centroid position of the initial state space robot system is selected as the origin O of the inertial system. Therefore, by substituting Eq. (31) into Eq. (35), the vector expression of the carrier centroid position can be obtained:

$$\mathbf{r}_0 = L_{00}\mathbf{e}_0 + L_{01}\mathbf{e}_1 + L_{02}\mathbf{e}_2 \tag{11}$$

Where $L_{00} = -(M - m_0)l_0/M$, $L_{01} = -(m_1a_1 + m_2l_1 + m_3l_1)/M$, $L_{02} = -(m_2a_2 + m_3l_2)/M$.

Suppose that the initial momentum of the system is zero, that is $\dot{\mathbf{r}}_c = 0$, from the conservation of momentum:

$$M\dot{\mathbf{r}}_c = \sum_{i=0}^{3} m_i \dot{\mathbf{r}}_i = 0 \tag{12}$$

The relationship between the velocity of point P at the end of the space robot arm in the inertial system and the generalized velocity $\dot{\mathbf{q}}$ is deduced as follows:

$$\mathbf{r}_p = \begin{bmatrix} \dot{x}_p \\ \dot{y}_p \end{bmatrix} = \begin{bmatrix} J_{11} & J_{12} & J_{13} \\ J_{21} & J_{22} & J_{23} \end{bmatrix} \begin{bmatrix} \dot{q}_0 \\ \dot{q}_1 \\ \dot{q}_2 \end{bmatrix} = \mathbf{J}\dot{\mathbf{q}} \tag{13}$$

where $\dot{\mathbf{q}} = \begin{pmatrix} \dot{q}_0 & \dot{q}_1 & \dot{q}_2 \end{pmatrix}^{\mathrm{T}}$, $\mathbf{J}$ is the Jacobian matrix of system velocity controlling carrier attitude. The items are as follows:

$J_{11} = L_{30}\cos q_0 + L_{31}\cos(q_0 + q_1) + L_{32}\cos(q_0 + q_1 + q_2)$,

$J_{12} = L_{31}\cos(q_0 + q_1) + L_{32}\cos(q_0 + q_1 + q_2), J_{13} = L_{32}\cos(q_0 + q_1 + q_2)$;

$J_{21} = -L_{30}\sin q_0 - L_{31}\sin(q_0 + q_1) - L_{32}\sin(q_0 + q_1 + q_2)$,

$J_{22} = -L_{31}\sin(q_0 + q_1) - L_{32}\sin(q_0 + q_1 + q_2), J_{23} = -L_{32}\sin(q_0 + q_1 + q_2)$

where $L_{10} = L_{00} + l_0$, $L_{11} = L_{01} + a_1$, $L_{12} = L_{02}$; $L_{20} = L_{00} + a_0$, $L_{21} = L_{01} + l_1$, $L_{22} = L_{02} + a_2$; $L_{30} = L_{00} + a_0$, $L_{31} = L_{01} + l_1$, $L_{32} = L_{02} + l_2$.

In this section, the deduced velocity Jacobian matrix $\mathbf{J}$ of the space robot system with no carrier position control and attitude control establishes the basis for the discussion of tracking the inertial space trajectory of the manipulator end of the space robot system in the following chapters.

The structure diagram of a fully rigid space robot system that does not control the position of the carrier but controls the attitude of the carrier is shown in **Figure 1**. Based on the velocity-vector relationship and the kinematics Jacobian matrix derived in Section 1.1, the dynamics equation of the system will be established in this section by combining the Lagrange equation of the second kind and the momentum conservation relationship.

According to the multi-body dynamics theory, according to aforesaid Eq. (13), the kinetic energy of each part of the space robot can be expressed as follows:

$$T_i = \frac{1}{2} m_i \dot{\mathbf{r}}_i^{\,2} + \frac{1}{2} I_i \omega_i^{\,2} \tag{14}$$

Where $i = 0, 1, 2, 3$, $\omega_0 = \dot{q}_0 \mathbf{z}_0$, $\omega_1 = (\dot{q}_0 + \dot{q}_1)\mathbf{z}_1$, $\omega_2 = (\dot{q}_0 + \dot{q}_1 + \dot{q}_2)\mathbf{z}_2$.

The kinetic energy of the system is expressed as follows:

$$\begin{aligned} T &= T_0 + T_1 + T_2 + T_3 = \frac{1}{2}\sum_{i=0}^{3}\left(m_i \dot{\mathbf{r}}_i^{\,2} + I_i \omega_i^{\,2}\right) \\ &= \frac{1}{2}\phi_1 W_1(\mathbf{q}, \dot{\mathbf{q}}) + \frac{1}{2}\phi_2 W_2(\mathbf{q}, \dot{\mathbf{q}}) + \frac{1}{2}\phi_3 W_3(\mathbf{q}, \dot{\mathbf{q}}) + \phi_4 W_4(\mathbf{q}, \dot{\mathbf{q}}) + \phi_5 W_5(\mathbf{q}, \dot{\mathbf{q}}) \\ &\quad + \phi_6 W_6(\mathbf{q}, \dot{\mathbf{q}}) \end{aligned} \tag{15}$$

Where $W_1 = \dot{q}_0{}^2$, $W_2 = (\dot{q}_0 + \dot{q}_1)^2$, $W_3 = (\dot{q}_0 + \dot{q}_1 + \dot{q}_3)^2$, $W_4 = \dot{q}_0(\dot{q}_0 + \dot{q}_1)\cos q_1$, $W_5 = \dot{q}_0(\dot{q}_0 + \dot{q}_1 + \dot{q}_2)\cos(q_1 + q_2)$, $W_6 = (\dot{q}_0 + \dot{q}_1)(\dot{q}_0 + \dot{q}_1 + \dot{q}_2)\cos q_2$, $\phi_i(i = 1 - 6)$ is the inertial parameter term, the specific items are as follows:

$$\phi_1 = m_0 L_{00}{}^2 + m_1 L_{10}{}^2 + m_2 L_{20}{}^2 + m_3 L_{30}{}^2 + I_0$$
$$\phi_2 = m_0 L_{01}{}^2 + m_1 L_{11}{}^2 + m_2 L_{21}{}^2 + m_3 L_{31}{}^2 + I_1$$
$$\phi_3 = m_0 L_{02}{}^2 + m_1 L_{12}{}^2 + m_2 L_{22}{}^2 + m_3 L_{32}{}^2 + I_2 + I_3$$
$$\phi_4 = m_0 L_{00} L_{01} + m_1 L_{10} L_{11} + m_2 L_{20} L_{21} + m_3 L_{30} L_{31}$$
$$\phi_5 = m_0 L_{00} L_{02} + m_1 L_{10} L_{12} + m_2 L_{20} L_{22} + m_3 L_{30} L_{32}$$
$$\phi_6 = m_0 L_{01} L_{02} + m_1 L_{11} L_{12} + m_2 L_{21} L_{22} + m_3 L_{31} L_{32}$$

If the weak gravitational factor is ignored and the system's gravitational potential energy $V = 0$ is ignored, the system's electromotive force is $L = T - V = T$. The space robot system takes the generalized coordinates $\mathbf{q} = \begin{pmatrix} q_0 & q_1 & q_2 \end{pmatrix}^{\mathrm{T}}$ and puts the Lagrange function of the system's dynamic force into the second Lagrange equation L as shown below:

$$\frac{\mathrm{d}}{\mathrm{d}t}\left(\frac{\partial L}{\partial \dot{\mathbf{q}}}\right) - \frac{\partial L}{\partial \mathbf{q}} = \mathbf{Q}_0 \tag{16}$$

Where $\mathbf{Q}_0$ is the generalized force vector. As shown in **Figure 1**, the dynamics equation of the space robot system is as follows:

$$\mathbf{D}(\mathbf{q})\ddot{\mathbf{q}} + \mathbf{H}(\mathbf{q}, \dot{\mathbf{q}})\dot{\mathbf{q}} = \boldsymbol{\tau} \tag{17}$$

Where $\mathbf{D}(\mathbf{q}) \in \mathbf{R}^{3\times3}$ is a symmetric, positive definite mass matrix, $\mathbf{H}(\mathbf{q}, \dot{\mathbf{q}})\dot{\mathbf{q}} \in \mathbf{R}^{3\times1}$ is a vector containing Coriolis force and centrifugal force. $\mathbf{q} = \begin{bmatrix} q_0 & q_1 & q_2 \end{bmatrix}^{\mathrm{T}}$ is the column vector of the generalized coordinates of the system. q_0 is the carrier attitude Angle, q_1, q_2 is the joint Angle of the boom, $\boldsymbol{\tau} \in \mathbf{R}^{3\times1}$ is the vector formed by the carrier attitude control moment and joint control moment.

The terms of the matrix $\mathbf{H}(\mathbf{q}, \dot{\mathbf{q}})$ are not unique, which can be properly selected $\mathbf{D}(\mathbf{q})$ and $\mathbf{H}(\mathbf{q}, \dot{\mathbf{q}})$ distinguished as follows:

$D_{11} = \phi_1 + \phi_2 + \phi_3 + 2\phi_4 \cos q_1 + 2\phi_5 \cos(q_1 + q_2) + 2\phi_6 \cos q_2$,

$D_{12} = \phi_2 + \phi_3 + \phi_4 \cos q_1 + \phi_5 \cos(q_1 + q_2) + 2\phi_6 \cos q_2$, $D_{13} = \phi_3 + \phi_5 \cos(q_1 + q_2) + \phi_6 \cos q_2$

$D_{21} = D_{12}$, $D_{22} = \phi_2 + \phi_3 + 2\phi_6 \cos q_2$, $D_{23} = \phi_3 + \phi_6 \cos q_2$, $D_{31} = D_{13}$, $D_{32} = D_{23}$, $D_{33} = \phi_3$.

$H_{11} = -\phi_4 \dot{q}_1 \sin q_1 - \phi_5(\dot{q}_1 + \dot{q}_2)\sin(q_1 + q_2) - \phi_6 \dot{q}_2 \sin q_2$,

$H_{12} = -\phi_4(\dot{q}_1 + \dot{q}_2)\sin q_1 - \phi_5(\dot{q}_0 + \dot{q}_1 + \dot{q}_2)\sin(q_1 + q_2) - \phi_6 \dot{q}_2 \sin q_2$,

$H_{13} = -\phi_5(\dot{q}_0 + \dot{q}_1 + \dot{q}_2)\sin(q_1 + q_2) - \phi_6(\dot{q}_0 + \dot{q}_1 + \dot{q}_2)\sin q_2$,

$H_{21} = \phi_4 \dot{q}_0 \sin q_1 + \phi_5 \dot{q}_0 \sin(q_1 + q_2) - \phi_6 \dot{q}_2 \sin q_2$,

$H_{22} = -\phi_6 \dot{q}_2 \sin q_2 H_{23} = -\phi_6(\dot{q}_0 + \dot{q}_1 + \dot{q}_2)\sin q_2$,

$H_{31} = \phi_5 \dot{q}_0 \sin(q_1 + q_2) + \phi_6(\dot{q}_0 + \dot{q}_1)\sin q_2$,

$H_{32} = \phi_6(\dot{q}_0 + \dot{q}_1)\sin q_2$, $H_{33} = 0$.

Then, $\mathbf{D}(\mathbf{q})$ and $\mathbf{H}(\mathbf{q}, \dot{\mathbf{q}})$ matrix satisfies the following relation:

$$\mathbf{z}^{\mathrm{T}}\mathbf{H}\mathbf{z} = \frac{1}{2}\mathbf{z}^{\mathrm{T}}\dot{\mathbf{D}}\mathbf{z} \tag{18}$$

Where $\mathbf{z} \in \mathrm{R}^{3\times1}$is any vector.

2.2 Kinematic modeling of flexible joint floating-based space robot system

As shown in **Figure 3**, the research object is a plane-moving flexible joint space robot. The space robot system consists of a floating carrier $B0$, two flexible joints and rigid arms $B1$, $B2$. The dynamics model of space robot system established in this section is aimed at the situation that the position of the carrier is not controlled but the attitude of the carrier is controlled.

The symbol definitions involved in formula derivation in **Figure 3** and this section is supplemented as follows:

K: represents the diagonal stiffness matrix of the joint of the manipulator;

q_i: actual joint Angle of the ith arm $(i = 1, 2)$;

q_{mi}: respectively refers to the Angle of the motor of the ith joint $(i = 1, 2)$;

ωi: the rotation angular speed of the carrier and each arm $(i = 0, 1, 2)$.

As shown in **Figure 4**, the joint of mechanical arm is generally composed of servo motor, harmonic reducer and arm rod. According to the flexible joint manipulator model studied by Spong [6], the joint flexibility can be simplified into a subsystem composed of joint actuator, linear torsional spring with constant inertia coefficient and arm.

Where K is the equivalent joint stiffness.

As shown in **Figure 4**, the floating space robot system satisfies the law of conservation of momentum without loss of generality. Assuming that the initial momentum of the system acting without external force is 0, the following formula can be obtained:

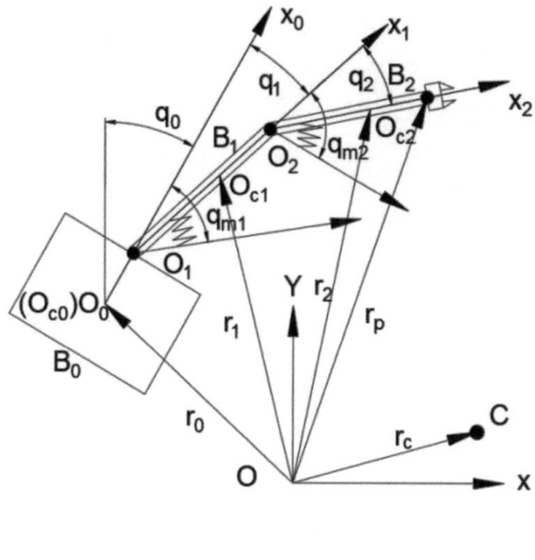

Figure 3.
Flexible-joint space robot.

$$M\dot{\mathbf{r}}_c = \sum_{i=0}^{3} m_i \dot{\mathbf{r}}_i = 0 \tag{19}$$

By taking the derivative of time t and combining with the total centroid relation (19), the expression of centroid velocity vector of each part can be obtained:

$$\dot{\mathbf{r}}_0 = \begin{bmatrix} -\dot{q}_0 L_{00} \sin q_0 - (\dot{q}_0 + \dot{q}_1) L_{01} \sin(q_0 + q_1) - (\dot{q}_0 + \dot{q}_1 + \dot{q}_2) L_{02} \sin(q_0 + q_1 + q_2) \\ \dot{q}_0 L_{00} \cos q_0 + (\dot{q}_0 + \dot{q}_1) L_{01} \cos(q_0 + q_1) + (\dot{q}_0 + \dot{q}_1 + \dot{q}_2) L_{02} \cos(q_0 + q_1 + q_2) \end{bmatrix} \tag{20}$$

$$\dot{\mathbf{r}}_1 = \begin{bmatrix} -\dot{q}_0 L_{10} \sin q_0 - (\dot{q}_0 + \dot{q}_1) L_{11} \sin(q_0 + q_1) - (\dot{q}_0 + \dot{q}_1 + \dot{q}_2) L_{12} \sin(q_0 + q_1 + q_2) \\ \dot{q}_0 L_{10} \cos q_0 + (\dot{q}_0 + \dot{q}_1) L_{11} \cos(q_0 + q_1) + (\dot{q}_0 + \dot{q}_1 + \dot{q}_2) L_{12} \cos(q_0 + q_1 + q_2) \end{bmatrix} \tag{21}$$

$$\dot{\mathbf{r}}_2 = \begin{bmatrix} -\dot{q}_0 L_{20} \sin q_0 - (\dot{q}_0 + \dot{q}_1) L_{21} \sin(q_0 + q_1) - (\dot{q}_0 + \dot{q}_1 + \dot{q}_2) L_{22} \sin(q_0 + q_1 + q_2) \\ \dot{q}_0 L_{20} \cos q_0 + (\dot{q}_0 + \dot{q}_1) L_{21} \cos(q_0 + q_1) + (\dot{q}_0 + \dot{q}_1 + \dot{q}_2) L_{22} \cos(q_0 + q_1 + q_2) \end{bmatrix} \tag{22}$$

$$\dot{\mathbf{r}}_p = \begin{bmatrix} -\dot{q}_0 L_{30} \sin q_0 - (\dot{q}_0 + \dot{q}_1) L_{31} \sin(q_0 + q_1) - (\dot{q}_0 + \dot{q}_1 + \dot{q}_2) L_{32} \sin(q_0 + q_1 + q_2) \\ \dot{q}_0 L_{30} \cos q_0 + (\dot{q}_0 + \dot{q}_1) L_{31} \cos(q_0 + q_1) + (\dot{q}_0 + \dot{q}_1 + \dot{q}_2) L_{32} \cos(q_0 + q_1 + q_2) \end{bmatrix} \tag{23}$$

The kinetic energy of each part of the system carrier and the boom is expressed as follows:

$$T_i = \frac{1}{2} m_i \dot{\mathbf{r}}_i^{\,2} + \frac{1}{2} I_i \omega_i^{\,2} \tag{24}$$

Where $i = 0, 1, 2, 3$, $\omega_0 = \dot{q}_0 \mathbf{z}_0$, $\omega_1 = (\dot{q}_0 + \dot{q}_1)\mathbf{z}_1$, $\omega_2 = (\dot{q}_0 + \dot{q}_1 + \dot{q}_2)\mathbf{z}_2$.
Kinetic energy of motor rotor component at the ith joint:

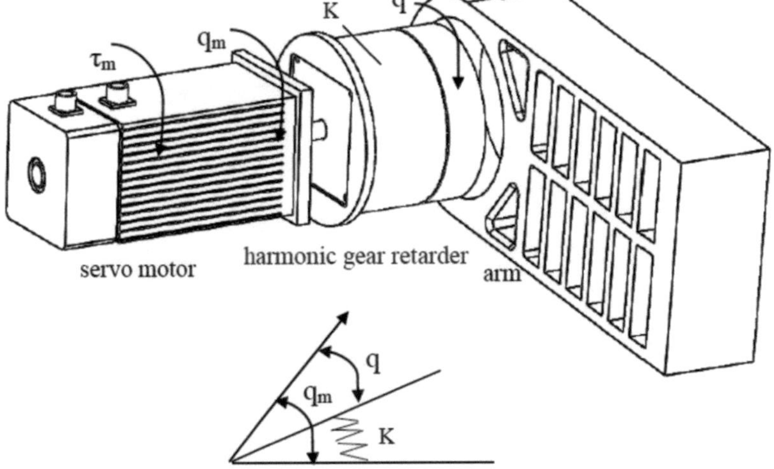

Figure 4.
Simplified model of flexible joint.

$$T_{\text{mi}} = \frac{1}{2} I_{im} \dot{q}_{mi}{}^{2}, (i = 1, 2) \tag{25}$$

Potential energy accumulated by elastic deformation at the *i*th joint:

$$V_{\text{mi}} = \frac{1}{2} k_{im} \left(\dot{q}_{mi} - \dot{q}_{i}\right)^{2}, (i = 1, 2) \tag{26}$$

The Lagrange emotive force of the system can be expressed as follows:

$$\begin{aligned} L &= T - V \\ &= \frac{1}{2}\sum_{i=0}^{3}\left(m_i \dot{\mathbf{r}}_i{}^2 + I_i \omega_i{}^2\right) + \sum_{i=1}^{2} T_{\text{mi}} - \sum_{i=1}^{2} V_{\text{mi}} \end{aligned} \tag{27}$$

According to the hypothesis (20)-(23), the space robot system satisfies the law of conservation of momentum. From the second Lagrange equation, the system dynamics equation of this kind of space robot can be obtained, which controls the attitude of the carrier but not the position of the carrier, without external force and without considering the weak gravity factor:

$$\mathbf{D}(\mathbf{q})\ddot{\mathbf{q}} + \mathbf{H}(\mathbf{q}, \dot{\mathbf{q}})\dot{\mathbf{q}} = \begin{bmatrix} \boldsymbol{\tau}_{\mathbf{0}} & \boldsymbol{\tau} \end{bmatrix}^{\mathrm{T}} \tag{28}$$

$$\mathbf{I}\ddot{\mathbf{q}}_{\mathbf{m}} + \boldsymbol{\tau} = \boldsymbol{\tau}_{\mathbf{m}} \tag{29}$$

Where $\ddot{\mathbf{q}}_{\mathbf{m}} = \left(\ddot{q}_{m1} \quad \ddot{q}_{m2}\right)^{\mathrm{T}}$, $\boldsymbol{\theta} = \begin{bmatrix} q_1 & q_2 \end{bmatrix}^{\mathrm{T}}$ are the actual response Angle of the two boom bars, the diagonal stiffness matrix $\mathbf{K} = \mathrm{diag}(k_1 \quad k_2) \in \mathrm{R}^{2\times 2}$, and $\boldsymbol{\tau}_{\mathbf{m}} \in \mathrm{R}^{2\times 1}$ is the control torque vector actually generated by the servo motor at the joint. And terms$\mathbf{D}(\mathbf{q})$ and $\mathbf{H}(\mathbf{q}, \dot{\mathbf{q}})$ are the same as in Eq. (18).

Similarly, the easy proof formulae (28) and (29) always satisfy the following formula:

$$\boldsymbol{\zeta}^{T} \boldsymbol{H} \boldsymbol{\zeta} = \frac{1}{2} \boldsymbol{\zeta}^{\mathrm{T}} \dot{\boldsymbol{D}} \boldsymbol{\zeta}$$

Where $\boldsymbol{\zeta} \in \mathrm{R}^{3\times 1}$ is any vector.

For the purpose of controller design, Eqs. (28) and (29) is decomposed as follows:

$$\mathbf{D}(\mathbf{q})\ddot{\mathbf{q}} + \mathbf{H}(\mathbf{q}, \dot{\mathbf{q}})\dot{\mathbf{q}} = \mathbf{W}\boldsymbol{\Phi} \tag{30}$$

Where $\boldsymbol{\Phi} = (\phi_1 \quad \phi_2 \quad \phi_3 \quad \phi_4 \quad \phi_5 \quad \phi_6)^{\mathrm{T}}$ is the combination vector of inertial parameters, $\mathbf{W}$ is the function matrix containing $\mathbf{q}$, $\dot{\mathbf{q}}$, and $\ddot{\mathbf{q}}$, all items do not contain inertial parameters, and the specific items are shown as follows:

$$\mathbf{W} = \begin{bmatrix} W_{11} & W_{12} & W_{13} & W_{14} & W_{15} & W_{16} \\ W_{21} & W_{22} & W_{23} & W_{24} & W_{25} & W_{26} \\ W_{31} & W_{32} & W_{33} & W_{34} & W_{35} & W_{36} \end{bmatrix}$$

Where $W_{11} = \ddot{q}_0$, $W_{12} = \ddot{q}_0 + \ddot{q}_1$, $W_{13} = \ddot{q}_0 + \ddot{q}_1 + \ddot{q}_2$, $W_{14} = \left(2\ddot{q}_0 + \ddot{q}_1\right)\cos q_1 -$ $\left(2\dot{q}_0 \dot{q}_1 + \dot{q}_1{}^2\right)\sin q_1$,

$$W_{15} = (2\ddot{q}_0 + \ddot{q}_1 + \ddot{q}_2)\cos(q_1 + q_2) - (2\dot{q}_0\dot{q}_1 + 2\dot{q}_0\dot{q}_2 + 2\dot{q}_1\dot{q}_2 + \dot{q}_1{}^2 + \dot{q}_2{}^2)\sin(q_1 + q_2),$$

$$W_{16} = (2\ddot{q}_0 + 2\ddot{q}_1 + \ddot{q}_2)\cos q_2 - (2\dot{q}_0\dot{q}_2 + 2\dot{q}_1\dot{q}_2 + \dot{q}_2{}^2)\sin q_2, W_{21} = 0,$$
$$W_{22} = W_{12}, W_{23} = W_{13},$$
$$W_{24} = \ddot{q}_0\cos q_1 + \dot{q}_0{}^2\sin q_1, W_{25} = \ddot{q}_0\cos(q_1 + q_2) + \dot{q}_0{}^2\sin(q_1 + q_2), W_{26} = W_{16},$$
$$W_{31} = 0, W_{32} = 0, W_{33} = W_{13}, W_{34} = 0, W_{35} = W_{25},$$
$$W_{36} = (\ddot{q}_0 + \ddot{q}_1)\cos q_2 - (\dot{q}_0 + \dot{q}_1)^2\sin q_2.$$

3. Robust adaptive control of space robot considering friction characteristics

The joint friction of space robot is complicated and nonlinear. The friction torque is the internal force of the system, and its coupling effect will affect the steady-state tracking error of the system. Conventional PID or robust method for trajectory tracking control has large steady-state error or limit cycle oscillation, and crawling occurs at low speed, which is difficult to achieve the expected real-time trajectory tracking effect [7, 8, 10]. At present, most control schemes considering frgziction use observer or neural network, which requires a large amount of computation and is not conducive to real-time online control of space robots [11].

In this chapter, a robust adaptive control strategy based on friction compensation is proposed by combining robust control theory and adaptive thought. The Lyapunov function of the system was constructed, and the tracking error convergence was proved and verified by numerical simulation. The proposed control strategy can effectively offset the influence of friction torque, improve the flat roof phenomenon in real-time tracking control, and improve the real-time trajectory tracking performance of the space robot.

As shown in **Figure 1**, based on the assumption that the attitude of the carrier is controlled and its position is not controlled, the dynamics equation of the fully rigid two-bar space robot system can be expressed as follows by reference in Eq. (17):

$$\mathbf{D}(\mathbf{q})\ddot{\mathbf{q}} + \mathbf{H}(\mathbf{q}, \dot{\mathbf{q}})\dot{\mathbf{q}} = \boldsymbol{\tau} \tag{31}$$

According to the analysis in Section 1, Eq. (30) shows that the dynamics equation of space robot can be expressed in the following form:

$$\mathbf{D}(\mathbf{q})\ddot{\mathbf{q}} + \mathbf{H}(\mathbf{q}, \dot{\mathbf{q}})\dot{\mathbf{q}} = \mathbf{W}\boldsymbol{\Phi} \tag{32}$$

Where $\boldsymbol{\Phi}$ is the combination vector of the inertial parameters, and $\mathbf{W}$ is the function matrix containing $\mathbf{q}$, $\dot{\mathbf{q}}$ and $\ddot{\mathbf{q}}$.

To represent the nominal value $\boldsymbol{\Phi}_0$ of the inertia parameter vector of the system and the error $\boldsymbol{\Phi}_e$ between the nominal value and the actual value of the inertia parameter vector, then:

$$\boldsymbol{\Phi} = \boldsymbol{\Phi}_0 + \boldsymbol{\Phi}_e \tag{33}$$

The space robot dynamics equation of the nominal model can be expressed as follows:

$$\mathbf{D}_0(\mathbf{q})\ddot{\mathbf{q}} + \mathbf{H}_0(\mathbf{q}, \dot{\mathbf{q}})\dot{\mathbf{q}} = \mathbf{W}\mathbf{\Phi}_0 \tag{34}$$

So the dynamics equation of the actual space robot system can be expressed as follows:

$$\mathbf{D}_0(\mathbf{q})\ddot{\mathbf{q}} + \mathbf{H}_0(\mathbf{q}, \dot{\mathbf{q}})\dot{\mathbf{q}} + \Delta\mathbf{H}(\mathbf{q}, \dot{\mathbf{q}}) = \boldsymbol{\tau} \tag{35}$$

Where $\Delta\mathbf{H}(\mathbf{q}, \dot{\mathbf{q}})$ represents the uncertainty caused by the inaccuracy of system structure parameters.

Considering the influence of joint friction factors on the system, the joint friction has strong nonlinear characteristics. When the joint rotation speed approaches zero, the friction torque at the joint is large. When the joint velocity is not zero, the friction torque is small, and the friction characteristics of the joint can be approximated by the nonlinear function shown in **Figure 2**.

It is difficult to accurately establish the specific function form $f(\dot{q})$ of friction torque, but its upper bound $\hat{f}(\dot{q})$ is known and can be expressed in the following form:

$$\begin{gathered} \hat{f}(\dot{q}) = \mu_s(\dot{q})\lambda(\dot{q}) + \mu_m(\dot{q})(1 - \lambda(\dot{q})) \\ |\mu_s(\dot{q})| \le \gamma_s, |\mu_m(\dot{q})| \le \gamma_f \end{gathered} \tag{36}$$

Where μ_m, μ_s represents the unknown static friction force and dynamic friction force, and γ_f, γ_s represents the upper bound of static friction and dynamic friction, respectively, and the switching function $\lambda(\dot{q})$ is expressed as follows:

$$\lambda(\dot{q}_i) = \begin{cases} 1, & |\dot{q}_i| > \boldsymbol{\varepsilon} \\ 0 & |\dot{q}_i| \le \boldsymbol{\varepsilon} \end{cases}, i = 1, 2 \tag{37}$$

Where $\boldsymbol{\varepsilon} > \boldsymbol{0}$ and sufficiently small.

The dynamics equation of the actual space robot model considering friction characteristics can be expressed as follows:

$$\mathbf{D}_0(\mathbf{q})\ddot{\mathbf{q}} + \mathbf{H}_0(\mathbf{q}, \dot{\mathbf{q}})\dot{\mathbf{q}} + \Delta\mathbf{H}(\mathbf{q}, \dot{\mathbf{q}}) = \boldsymbol{\tau} - \mathbf{f} \tag{38}$$

Where f represents the friction torque, and the rotation of the carrier relative to the inertial system is not affected by the friction torque, $f(1) = 0$.

Let $\mathbf{q}_d$ be the expected trajectory of the system, the design control law is as follows:

$$\boldsymbol{\tau} = \mathbf{u} + \mathbf{D}_0(\mathbf{q})\ddot{\mathbf{q}}_d + \mathbf{H}_0(\mathbf{q}, \dot{\mathbf{q}})\dot{\mathbf{q}}_d \tag{39}$$

Joint tracking error $\mathbf{e} = \mathbf{q} - \mathbf{q}_d$ is defined. The actual space robot system error equation can be expressed as follows:

$$\mathbf{D}_0(\mathbf{q})\ddot{e} + \mathbf{H}_0(\mathbf{q}, \dot{\mathbf{q}})\dot{e} + \mathbf{\Delta H} = u - f \tag{40}$$

Define state variables as follows:

$$\begin{cases} \mathbf{x}_1 = \mathbf{e} \\ \mathbf{x}_2 = \dot{\mathbf{e}} + \mathbf{e} \end{cases} \tag{41}$$

The state equation of system error can be obtained from Eqs. (38) and (41) as follows:

$$\begin{cases} \dot{\mathbf{x}}_1 = \mathbf{x}_2 - \mathbf{x}_1 \\ \mathbf{D}_0\dot{\mathbf{x}}_2 = \mathbf{D}_0(\mathbf{x}_2 - \mathbf{x}_1) - \mathbf{H}_0(\mathbf{x}_2 - \mathbf{x}_1) + \mathbf{u} - \mathbf{f} - \Delta\mathbf{H} \end{cases} \tag{42}$$

The design control law $\mathbf{u}$ is as follows:

$$\mathbf{u} = \mathbf{u}_{\mathrm{r}} + \mathbf{u}_{\mathrm{c}} + \mathbf{u}_{\mathrm{f}} \tag{43}$$

Where $\mathbf{u}_{\mathrm{r}}$ is the error compensation term, $\mathbf{u}_{\mathrm{c}}$ represents the structural uncertainty fitting term, and $\mathbf{u}_{\mathrm{f}}$ is the friction compensation term. The specific form is as follows:

$$\begin{cases} \mathbf{u}_{\mathrm{r}} = -\mathbf{D}_0(\mathbf{x}_2 - \mathbf{x}_1) - \mathbf{H}_0\mathbf{x}_1 - \mathbf{x}_1 - p_2\mathbf{x}_2 + \dfrac{1}{4\gamma^2}\mathbf{x}_2 \\ \mathbf{u}_c = W\hat{\mathbf{\Phi}}_{\mathrm{e}} \\ \mathbf{u}_f = -\boldsymbol{\lambda}(\dot{\mathbf{q}})\,\mathrm{sgn}(\mathbf{x}_2)\gamma_s - \left(1 - \boldsymbol{\lambda}(\dot{\mathbf{q}})\,\mathrm{sgn}(\mathbf{x}_2)\gamma_f\right) \end{cases} \tag{44}$$

Where $\hat{\mathbf{\Phi}}_{\mathrm{e}}$ represents the estimated error of the nominal value $\mathbf{\Phi}_e$ and the actual value of the inertial parameter vector, which P_2 is greater than zero and is a positive constant.

The uncertainty caused by the inaccuracy of system structure parameters $\Delta\mathbf{H}(\mathbf{q}, \dot{\mathbf{q}})$ is compensated by the fitting merge $\mathbf{u}_{\mathrm{c}}$:

$$\Delta\mathbf{H}(\mathbf{q}, \dot{\mathbf{q}}) = \mathbf{W}\hat{\mathbf{\Phi}}_{\mathrm{e}} + \boldsymbol{\varepsilon}_{\mathbf{f}} \tag{45}$$

Where $\boldsymbol{\varepsilon}_{\mathbf{f}}$ is the fitting error of structural uncertainty term.

The law of design adaptation is as follows:

$$\dot{\hat{\mathbf{\Phi}}}_{\mathrm{e}} = -\lambda\mathbf{W}^{\mathrm{T}}\mathbf{x}_2 \tag{46}$$

Where $\lambda > 0$ is a constant. Defined estimation error $\tilde{\mathbf{\Phi}}_e = \hat{\mathbf{\Phi}}_{\mathrm{e}} - \mathbf{\Phi}_e$.

The evaluation signal $\mathbf{Z} = P_1\mathbf{x}_1$ is defined, and the fitting error $\boldsymbol{\varepsilon}_{\mathbf{f}}$ of the uncertain term is regarded as the perturbation term of the closed-loop system. To prove that the closed-loop system satisfies L2 gain condition, the Lyapunov-like function of the system is constructed, which makes the dissipation inequality as shown below valid:

$$\dot{V} \le \left\{\gamma^2\|\boldsymbol{\varepsilon}_{\mathbf{f}}\|^2 - \|\mathbf{z}(t)\|^2\right\} \tag{47}$$

Where $\gamma > 0$ represents the interference suppression level.

Construct the Lyapunov function as follows:

$$V = \frac{1}{2}{\mathbf{x}_1}^T\mathbf{x}_1 + \frac{1}{2}{\mathbf{x}_2}^T\mathbf{D}_0\mathbf{x}_2 + \frac{1}{2\lambda}\mathrm{tr}\left({\tilde{\mathbf{\Phi}}_e}^{\mathrm{T}}\tilde{\mathbf{\Phi}}_e\right) \tag{48}$$

By taking the derivative of the above equation with respect to time, we can obtain from the dissipation inequality (47):

$$
\begin{aligned}
H &= \dot{V} - \left(\gamma^2\|\boldsymbol{\varepsilon}_{\mathrm{f}}\|^2 - \|\mathbf{z}(t)\|^2\right) \\
&= \mathbf{x}_1{}^T\dot{\mathbf{x}}_1 + \mathbf{x}_2{}^T\left(\mathbf{D}_0\dot{\mathbf{x}}_2 + \frac{1}{2}\dot{\mathbf{D}}_0\mathbf{x}_2\right) - \mathbf{W}^T\hat{\boldsymbol{\Phi}}_{\mathrm{e}}\mathbf{x}_2 - \left(\gamma^2\|\boldsymbol{\varepsilon}_{\mathrm{f}}\|^2 - \|\mathbf{z}(t)\|^2\right) \\
&= \left(1 - \mathrm{P}_1{}^2\right)\|\mathbf{x}_1\|^2 - \mathbf{x}_2{}^{\mathrm{T}}\boldsymbol{\varepsilon}_{\mathrm{f}} - \frac{1}{\lambda}\mathrm{tr}\left(\dot{\tilde{\boldsymbol{\Phi}}}_e{}^{\mathrm{T}}\tilde{\boldsymbol{\Phi}}_e\right) - \gamma^2\|\boldsymbol{\varepsilon}_{\mathrm{f}}\|^2 \\
&+ \mathbf{x}_2{}^T\left(-P_2\mathbf{x}_2 + \mathbf{W}^T\hat{\boldsymbol{\Phi}}_{\mathrm{e}} + \frac{1}{4\gamma^2}\mathbf{x}_2 + \mathbf{u}_{\mathrm{f}} - \mathbf{f}\right) \\
&\le -\eta\|\mathbf{x}_1\|^2 - P_2\|\mathbf{x}_2\|^2 + \mathbf{x}_2{}^T(\mathbf{u}_{\mathrm{f}} - \mathbf{f})
\end{aligned}
\tag{49}
$$

Where $\eta < 1 - \mathrm{P}_1{}^2$ is a positive constant.

By analyzing the friction compensation term $\mathbf{u}_{\mathrm{f}}$, the following inequality relation is established:

$$
\mathbf{x}_2 > 0, \quad \begin{cases} -\gamma_s - \mu_s(\dot{q}) < 0, |\dot{q}| > \boldsymbol{\varepsilon} \\ -\gamma_f - \mu_m(\dot{q}) < 0, |\dot{q}| \le \boldsymbol{\varepsilon} \end{cases}
$$

$$
\mathbf{x}_2 < 0, \quad \begin{cases} \gamma_s - \mu_s(\dot{q}) > 0, |\dot{q}| > \boldsymbol{\varepsilon} \\ \gamma_f - \mu_m(\dot{q}) > 0, |\dot{q}| \le \boldsymbol{\varepsilon} \end{cases}
$$

Therefore, $\mathbf{x}_2{}^T(\mathbf{u}_{\mathrm{f}} - \mathbf{f}) < 0$ is true, $H < 0$ is always true, and the closed-loop system meets L2 gain condition $J < \gamma$. According to Lyapunov's stability theorem, the closed-loop system satisfies the asymptotic stability condition, that is, $\lim_{t\to\infty}\mathbf{e}(t) = 0, \lim_{t\to\infty}\dot{\mathbf{e}}(t) = 0$.

The research object of value simulation experiment is the planar two-bar space robot as shown in **Figure 1**.

The inertia parameters of the system are selected as follows:

Quality of each component $m_0 = 40\mathrm{kg}$, $m_1 = 2\mathrm{kg}$, $m_2 = 1\mathrm{kg}$; the moment of inertia are respectively $J_0 = 34.17\mathrm{kg}\cdot\mathrm{m}^2, J_1 = 1.5\mathrm{kg}\cdot\mathrm{m}^2, J_2 = 0.75\mathrm{kg}\cdot\mathrm{m}^2$; the mechanical arm is set as a uniform rod, and the length of the rod is respectively $l_1 = 3\mathrm{m}$, $l_2 = 3\mathrm{m}$, where the distance along the axis x_0 is $l_0 = 1.5\mathrm{m}$ from point $O\boldsymbol{0}$ to $O\boldsymbol{1}$. The unknown load at the end of the nominal arm $m_{\mathrm{p}} = 2\mathrm{kg}$, from the end load to the moment of inertia O_{c2} is $J_3 = 1.5\mathrm{kg}\cdot\mathrm{m}^2$. The actual load acting on the end of the manipulator is $m_{\mathrm{pe}} = 4\mathrm{kg}$, other parameters are the same as the nominal model. Initial configuration of system simulation $q_0(0) = 0.1\mathrm{rad}$, $q_1(0) = 0.22\mathrm{rad}$,$q_2(0) = 0.79\mathrm{rad}$ and expected trajectory are as follows:

$$
\begin{cases} q_{\mathrm{d}0} = \pi/6\mathrm{rad} \\ q_{\mathrm{d}1} = \pi/8\text{-}\sin(\pi t/3)\mathrm{rad} \\ q_{\mathrm{d}2} = \cos(\pi t/3)\mathrm{rad} \end{cases}
$$

The continuous smooth static friction law proposed by Feeny was adopted to represent the friction characteristics of joints [12]:

$$
\mathbf{f}(\dot{\mathbf{q}}) = \mathbf{f}_{\mathrm{v}}\dot{\mathbf{q}} + (\mathbf{f}_{\mathrm{c}} + \mathbf{f}_{\mathrm{cs}}\mathrm{sech}(\beta\dot{\mathbf{q}})\tanh(\alpha\dot{\mathbf{q}}))
$$

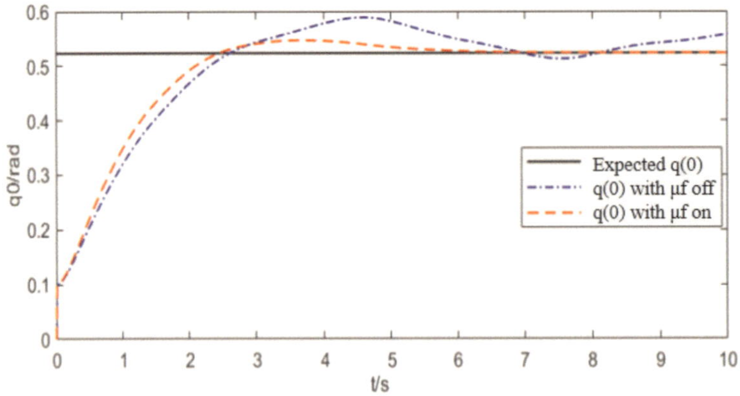

Figure 5.
Carrier attitude angle.

The viscous friction factor $f_v = 0.00986$; $f_c = 0.743$ is coulomb friction factor; $f_{cs} = 3.99$ and $\alpha = 3.24$ structure Stribeck friction; $\beta = 0.799$ ensure the continuity of the equation when the relative sliding velocity is zero, $f(1) = 0$.

The simulation results are as follows:

As can be seen in **Figure 5**, when friction compensation control is not enabled, the coupling effect of joint friction torque will cause interference to the attitude of the carrier, making it difficult to maintain the desired attitude Angle stably. **Figures 6** and **7** show the tracking response of the system to the command trajectory. Due to the existence of friction factors, the joint Angle tracking has obvious response hysteresis, and the actual trajectory lags behind the expected trajectory, that is, $|\dot{\mathbf{q}}| \leq |\dot{\mathbf{q}}_d|$. With the friction torque compensation control turned on, the simulation results show that the friction compensation term can effectively offset the coupling effect of joint friction torque on the carrier, and can maintain the desired attitude of the carrier stably. After friction compensation control, the hysteresis phenomenon of joint Angle tracking is effectively improved, and the real-time trajectory tracking performance is improved.

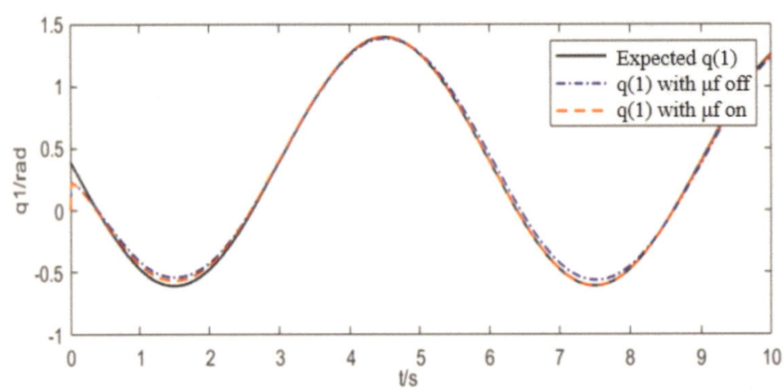

Figure 6.
Joint angle of arm 1.

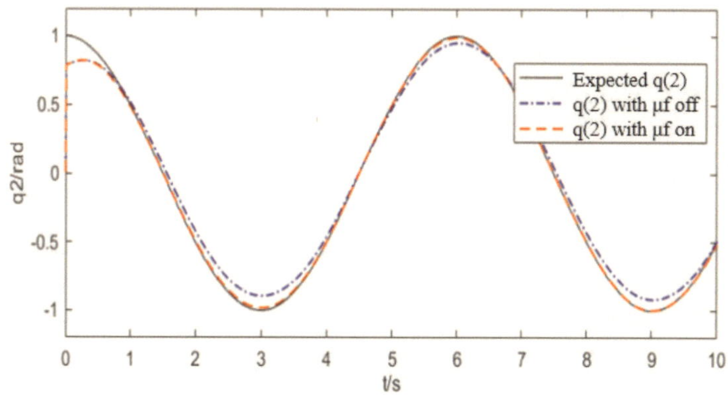

Figure 7.
Joint angle of arm 2.

3.1 Summary

1. This section discusses the trajectory tracking control problem of a fully rigid space robot with uncertain structural parameters under the influence of joint friction torque, and designs a robust adaptive control strategy based on friction compensation.

2. The Lyapunov function of the system was constructed, and the convergence of tracking errors was proved and verified by numerical simulation. The design scheme can effectively improve the response hysteria caused by friction factor, and effectively reduce the influence of joint coupling factors on the carrier attitude control.

3. The designed controller has the advantages of simple structure and small amount of calculation, and can be applied to real-time trajectory tracking control of three-dimensional space robots by extension.

4. Robust control and active suppression of elastic vibration of flexible joint space robot considering friction characteristics

In practical engineering applications, the space manipulator mostly uses harmonic reducer as the joint transmission mechanism. The application of harmonic reducer makes the joint of the manipulator have inevitable flexibility, which brings the coupling between the servo motor Angle and the joint Angle, and easily causes the manipulator jitter and the system response hysteretic problems [12–14]. In the undamped space environment, the vibration attenuation is slow, which brings great difficulty to the control of space robot. At present, there are few control schemes for space robots that take joint friction into consideration, and most studies take friction into consideration the disturbance term [7, 15–17]. In most control schemes, friction compensation is based on on-line observer, and the friction torque is observed and compensated as the disturbance term. Moreover, the structure of most controllers is complex or dependent on neural network, which is not conducive to real-time joint control [18–20].

In this section, a flexible joint space robot with uncontrolled carrier position and attitude is taken as the research object, and the control problem of space robot under the influence of strong nonlinear friction torque is explored. Based on the joint flexibility compensation idea and singular perturbation theory, a L2 gain robust controller based on the upper bound of friction force was designed, and the Lyapunov function of the system was constructed. The convergence of tracking errors was proved and verified by co-simulation.

As shown in **Figure 3**, the object of study is a space robot system with flexible joints that moves in a plane. The system whose attitude is controlled by the reaction wheel satisfies the law of conservation of momentum. Let the initial momentum be 0. Based on the assumption that the attitude of the carrier is controlled without controlling its position, the system dynamics model of the flexible joint space robot is derived from Eqs. (28) and (29):

$$\mathbf{D}(\mathbf{q})\ddot{\mathbf{q}} + \mathbf{H}(\mathbf{q}, \dot{\mathbf{q}})\dot{\mathbf{q}} = \begin{bmatrix} \boldsymbol{\tau}_0 & \boldsymbol{\tau} \end{bmatrix}^{\mathrm{T}} \tag{50}$$

$$\mathbf{J}\ddot{\boldsymbol{\theta}}_{\mathbf{m}} + \boldsymbol{\tau} = \boldsymbol{\tau}_{\mathbf{m}} \tag{51}$$

$$\boldsymbol{\tau} = \mathbf{K}(\boldsymbol{\theta}_{\mathbf{m}} - \boldsymbol{\theta}) \tag{52}$$

The rotational inertia $\mathbf{J} \in \mathrm{R}^{2\times 2}$ of the motor is the diagonal and constant matrix, and$\mathbf{K} \in \mathrm{R}^{2\times 2}$ is the diagonal stiffness matrix of the joint.
$\mathbf{q} = \begin{bmatrix} q_0 & q_1 & q_2 \end{bmatrix}^{\mathrm{T}}$,$\boldsymbol{\theta} = \begin{bmatrix} q_1 & q_2 \end{bmatrix}^{\mathrm{T}}$ is the actual rotation Angle of the boom, $\boldsymbol{\theta}_{\mathbf{m}} = \begin{bmatrix} q_{m1} & q_{m2} \end{bmatrix}^{\mathrm{T}}$ is the motor rotation Angle vector, $\boldsymbol{\tau} \in \mathrm{R}^{2\times 1}$ is the control torque vector acting on the mechanical arm, $\boldsymbol{\tau}_{\mathrm{m}} \in \mathrm{R}^{2\times 1}$ represents the actual control torque vector of the servo motor.

The dynamic equation of torque can be obtained from Eqs. (51) and (52):

$$\mathbf{J}\mathbf{K}^{-1}\ddot{\boldsymbol{\tau}} + \boldsymbol{\tau} = \boldsymbol{\tau}_{\mathrm{m}} - \mathbf{J}\ddot{\boldsymbol{\theta}} \tag{53}$$

Let the motor control input be as follows:

$$\boldsymbol{\tau}_{\mathrm{m}} = \mathbf{K}_{\mathrm{r}}\boldsymbol{\tau}_{\mathrm{r}} + \mathbf{u}_{\mathrm{fs}} \tag{54}$$

Where $\mathbf{K}_{\mathrm{r}} = \mathrm{I} + \mathbf{K}_{\mathrm{b}}$, $\mathbf{K}_{\mathrm{b}} \in \mathrm{R}^{2\times 2}$ is the diagonal, positive invariant matrix, I is the identity matrix; $\boldsymbol{\tau}_{\mathrm{r}} \in \mathrm{R}^{2\times 1}$ is the control quantity to be designed; Joint flexibility compensation controller $\mathbf{u}_{\mathrm{fs}} = -\mathbf{K}_{\mathrm{b}}\boldsymbol{\tau}$. According to Eqs. (54) and (53), it can be obtained that:

$$\mathbf{J}\mathbf{K}_{\mathrm{e}}^{-1}\ddot{\boldsymbol{\tau}} + \boldsymbol{\tau} = \boldsymbol{\tau}_{\mathrm{r}} - \mathbf{K}_{\mathrm{r}}^{-1}\mathbf{J}\ddot{\boldsymbol{\theta}} \tag{55}$$

Where $\mathbf{K}_{\mathrm{e}} = \mathbf{K}\mathbf{K}_{\mathrm{r}}$ is the equivalent stiffness matrix after flexibility compensation. The control input $\boldsymbol{\tau}_{\mathrm{r}}$ based on singular perturbation is designed as follows:

$$\boldsymbol{\tau}_{\mathrm{r}} = \boldsymbol{\tau}_{\mathrm{rl}} + \boldsymbol{\tau}_{\mathrm{rf}} \tag{56}$$

Where $\boldsymbol{\tau}_{\mathrm{rl}} \in \mathrm{R}^{2\times 1}$ is the control input of slow variable subsystem and $\boldsymbol{\tau}_{\mathrm{rf}} \in \mathrm{R}^{2\times 1}$ is the control input of fast variable subsystem.

Define sufficiently small positive numbers $\mu > 0$, let $\mathbf{K}_{\mathrm{e}} = \mathbf{K}_{\mathrm{a1}}/\mu^2$; $\mathbf{K}_{\mathrm{a1}} \in \mathrm{R}^{2\times 2}$ is of a similar order of magnitude to the variable in the slow subsystem. The fast subsystem equation is derived from Eq. (55) as follows:

$$\mu^2\mathbf{J}\ddot{\boldsymbol{\tau}} + \mathbf{K}_{a1}\boldsymbol{\tau} = \mathbf{K}_{a1}\left(\boldsymbol{\tau}_{rl} - \mathbf{K}_r^{-1}\mathbf{J}\ddot{\boldsymbol{\theta}}\right) + \mathbf{K}_{a1}\boldsymbol{\tau}_{rf} \tag{57}$$

The torque differential controller is designed as follows:

$$\boldsymbol{\tau}_{rf} = -\mu\mathbf{K}_{a2}\mathbf{K}\left(\dot{\boldsymbol{\theta}}_m - \dot{\boldsymbol{\theta}}\right) \tag{58}$$

Let $\mu = 0$, the dynamic equation of the slow subsystem is as follows:

$$\overline{\mathbf{D}}(\mathbf{q})\ddot{\mathbf{q}} + \mathbf{H}(\mathbf{q}, \dot{\mathbf{q}})\dot{\mathbf{q}} = \begin{bmatrix}\tau_0 & \boldsymbol{\tau}_{rl}\end{bmatrix}^T \tag{59}$$

Among them $\overline{\mathbf{D}}(\mathbf{q}) = \begin{bmatrix}\overline{\mathbf{D}}_{11} & \overline{\mathbf{D}}_{12} \\ \overline{\mathbf{D}}_{21} & \overline{\mathbf{D}}_{22}\end{bmatrix}$, $\overline{\mathbf{D}}_{11} = \mathbf{D}_{11}$, $\overline{\mathbf{D}}_{12} = \mathbf{D}_{12}$, $\overline{\mathbf{D}}_{21} = \mathbf{D}_{21}$, $\overline{\mathbf{D}}_{22} = \mathbf{D}_{22} + \mathbf{K}_r^{-1}\mathbf{J}$; where $\mathbf{K}_r^{-1}\mathbf{J}$ is a constant matrix, the first derivative is always zero, and Eq. (17) can be proved similarly. For any column vector $\mathbf{z} \in R^{3\times1}$, it is always:

$$\mathbf{z}^T\mathbf{H}\mathbf{z} = \frac{1}{2}\mathbf{z}^T\dot{\overline{\mathbf{D}}}\mathbf{z} \tag{60}$$

The dynamic equation of slow subsystem considering the influence of joint friction can be expressed as follows:

$$\overline{\mathbf{D}}(\mathbf{q})\ddot{\mathbf{q}} + \mathbf{H}(\mathbf{q}, \dot{\mathbf{q}})\dot{\mathbf{q}} = \begin{bmatrix}\tau_0 & \boldsymbol{\tau}_{rl}\end{bmatrix}^T - f \tag{61}$$

Where f represents the friction torque, $f(1) = 0$.

The robot joint friction shows strong nonlinear characteristics at low speed, as shown in **Figure 8**. When the joint rotation speed approaches zero, the friction torque at the joint is large. When the joint velocity is not zero, the friction torque is small. Many scholars have studied the friction characteristics of joint, and the precise friction model is complicated, which makes the calculation amount surge. Consider using the nonlinear function shown in **Figure 8** to approximate the friction characteristics of the joint.

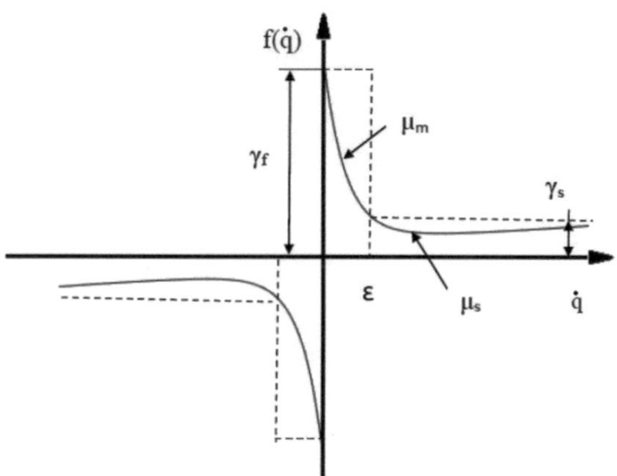

Figure 8.
Friction characteristics of joints.

In this section, it is studied that the form of joint friction characteristic function is difficult to accurately establish, but its upper bound $\hat{f}(\dot{q})$ is known, which is expressed as follows:

$$\begin{gathered}\hat{f}(\dot{q}) = \mu_s(\dot{q})\lambda(\dot{q}) + \mu_m(\dot{q})(1-\lambda(\dot{q})) \\ |\mu_s(\dot{q})| \leq \gamma_s, |\mu_m(\dot{q})| \leq \gamma_f\end{gathered} \tag{62}$$

Where μ_f, μ_s represent the unknown static friction force and dynamic friction force, γ_f and γ_s are the upper bound of static friction and dynamic friction, respectively, $\lambda(\dot{q})$ is the switching function, expressed as follows:

$$\lambda(\dot{q}_i) = \begin{cases} 1, & |\dot{q}_i| > \boldsymbol{\varepsilon} \\ 0 & |\dot{q}_i| \leq \boldsymbol{\varepsilon} \end{cases}, i = 1, 2 \tag{63}$$

Where $\boldsymbol{\varepsilon} > \boldsymbol{0}$ and sufficiently small.

Set $\mathbf{q}_d$ as the expected output of the system, and define the joint Angle tracking error as follows:

$$\mathbf{e} = \mathbf{q} - \mathbf{q}_d \tag{64}$$

By introducing the control quantity $\mathbf{u}$ to be designed, the space robot tracking problem is transformed into an asymptotic stability control problem:

$$\overline{\mathbf{D}}(\mathbf{q})\ddot{\mathbf{q}}_d + \mathbf{H}(\mathbf{q}, \dot{\mathbf{q}})\dot{\mathbf{q}}_d + \mathbf{u} = [\tau_0 \quad \boldsymbol{\tau}_{rl}]^T \tag{65}$$

Considering the external interference torque $\mathbf{d} \in R^{3\times1}$ acting on the system, the dynamics model of the slow subsystem is as follows:

$$\overline{\mathbf{D}}(\mathbf{q})\ddot{\mathbf{q}} + \mathbf{H}(\mathbf{q}, \dot{\mathbf{q}})\dot{\mathbf{q}} = [\tau_0 \quad \boldsymbol{\tau}_{rl}]^T + \mathbf{d} - f \tag{66}$$

Let $\mathbf{x}_1 = \mathbf{e}$, $\mathbf{x}_2 = \dot{\mathbf{e}} + \mathbf{e}$ and the state-space equation of the dynamics model of space robot system considering friction characteristics and external interference torque is as follows:

$$\begin{cases} \dot{\mathbf{x}}_1 = \mathbf{x}_2 - \mathbf{x}_1 \\ \overline{\mathbf{D}}\dot{\mathbf{x}}_2 = \overline{\mathbf{D}}(\mathbf{x}_2 - \mathbf{x}_1) - \mathbf{H}(\mathbf{x}_2 - \mathbf{x}_1) + \mathbf{u} + \mathbf{d} - \mathbf{f} \end{cases} \tag{67}$$

The controller $\mathbf{u}$ is designed as follows:

$$\mathbf{u} = \mathbf{u}_r + \mathbf{u}_d + \mathbf{u}_f \tag{68}$$

Where $\mathbf{u}_r$ is the tracking error compensation term, $\mathbf{u}_d$ is the disturbance compensation term and $\mathbf{u}_f$ is the friction compensation term. The evaluation signal $\mathbf{Z} = [P_1\mathbf{e}_1 \quad P_2\mathbf{e}_2]^T$, $P_1 \geq 0$, $P_2 \geq 0$ is defined as the given weighting coefficient, which is expressed as follows:

$$\begin{cases} \mathbf{u}_r = -\overline{\mathbf{D}}(\mathbf{x}_2 - \mathbf{x}_1) - \mathbf{H}\mathbf{x}_1 - \mathbf{x}_1 \\ \mathbf{u}_d = 2{P_2}^2\mathbf{x}_1 - \mathbf{K}_d\mathbf{x}_2 \\ \mathbf{u}_{fs} = -\left(\boldsymbol{\lambda}(\dot{\mathbf{q}})\,\mathrm{sgn}(\mathbf{x}_2)\gamma_s + \left(1 - \boldsymbol{\lambda}(\dot{\mathbf{q}})\,\mathrm{sgn}(\mathbf{x}_2)\gamma_f\right)\right) \end{cases} \tag{69}$$

Construct the energy function V. If the system satisfies L2 gain condition, then the following dissipation inequality holds:

$$\dot{V} \leq \left\{ \gamma^2 \|\mathbf{d}(t)\|^2 - \|\mathbf{z}(t)\|^2 \right\} \tag{70}$$

Where $\gamma > 0$ represents the suppression level of system interference, and the value γ reflects the influence of external disturbance on system tracking error.

The Lyapunov function of the closed-loop system is constructed as follows:

$$V = \frac{1}{2} \mathbf{x}_1{}^T \mathbf{x}_1 + \frac{1}{2} \mathbf{x}_2{}^T \overline{\mathbf{D}} \mathbf{x}_2 \tag{71}$$

The derivative of V with respect to time T is:

$$\begin{aligned} H &= \dot{V} - \left(\gamma^2 \|\mathbf{d}(t)\|^2 - \|\mathbf{z}(t)\|^2 \right) \\ &= \mathbf{x}_1{}^T \mathbf{x}_1 + P_1{}^2 \|\mathbf{x}_1\|^2 + P_2{}^2 \|\mathbf{x}_2 - \mathbf{x}_1\|^2 - \gamma^2 \|\mathbf{d}(t)\|^2 \\ &+ \mathbf{x}_2{}^T \left[\overline{\mathbf{D}}(\mathbf{x}_2 - \mathbf{x}_1) + \mathbf{H}\mathbf{x}_1 - \mathbf{x}_1 + \mathbf{u} + \mathbf{d} - \mathbf{f} \right] \\ &\leq -\delta_1 \|\mathbf{x}_1\|^2 - \delta_2 \|\mathbf{x}_2\|^2 + \mathbf{x}_2{}^{\mathrm{T}} (\mathbf{u}_{\mathrm{fs}} - \mathbf{f}) \end{aligned} \tag{72}$$

Where δ_1 and δ_2 are sufficiently small but greater than 0, and satisfies the following formula:

$$\begin{cases} 1 - P_1{}^2 - P_2{}^2 \geq \delta_1 \\ \|\mathbf{K}_{\mathrm{d}}\| - P_2{}^2 - \dfrac{1}{4\gamma^2} \geq \delta_2 \end{cases} \tag{73}$$

From Eq. (69), we can see that the following inequality relationship is valid:

$$\mathbf{x}_2 > 0, \quad \begin{cases} -\gamma_s - \mu_s(\dot{q}) < 0, |\dot{q}| > \varepsilon \\ -\gamma_f - \mu_m(\dot{q}) < 0, |\dot{q}| \leq \varepsilon \end{cases}$$
$$\mathbf{x}_2 < 0, \quad \begin{cases} \gamma_s - \mu_s(\dot{q}) > 0, |\dot{q}| > \varepsilon \\ \gamma_f - \mu_m(\dot{q}) > 0, |\dot{q}| \leq \varepsilon \end{cases}$$

Therefore, $H < 0$ is always true, and the closed-loop system meets L2 gain condition$J < \gamma$. According to Lyapunov's stability theorem, the closed-loop system satisfies the asymptotic stability condition, that is, $\lim_{t\to\infty} \mathbf{e}(t) = 0$, $\lim_{t\to\infty} \dot{\mathbf{e}}(t) = 0$.

The inertia parameters of the system are selected as follows:

Carrier parameters are $m_0 = 40\text{kg}$, $I_0 = 34.17\text{kg.m}^2$, $l_0 = 1.5\text{m}$; the inertia parameter of boom rod 1 is $m_1 = 2\text{kg}$, $I_1 = 1.5\text{kg.m}^2$, $l_1 = 3\text{m}$, and the inertia parameter of boom rod 2 is $m_2 = 1\text{kg}$, $I_2 = 0.8\text{kg.m}^2$, $l_2 = 3\text{m}$. The stiffness of the flexible joint is set $\mathbf{K} = \mathrm{diag}(k_1, k_2)$, where $k_1 = 100\text{N} \cdot \text{m/rad}$,$k_2 = 100\text{N} \cdot \text{m/rad}$; the moment of inertia of the motor $\mathbf{J} = \mathrm{diag}(J_1, J_2)$, where $J_1 = 0.08\text{kg} \cdot \text{m}^2$, $J_2 = 0.08\text{kg} \cdot \text{m}^2$. Select the controller parameter $P_1 = 0.5$, $P_2 = 0.5$, $\mathrm{K_d} = \boldsymbol{diag}(\mathbf{50}, \mathbf{50}, \mathbf{50})$, $\mathrm{K_b} = \boldsymbol{diag}(65, 65)$, $\gamma = 1$, $\gamma_{\mathrm{s}} = 2$, $\gamma_{\mathrm{f}} = 4$, $\mathbf{K}_{\mathrm{a1}} = \boldsymbol{diag}(\mathbf{0.05}, \mathbf{0.05})$, $\mathbf{K}_{\mathrm{a2}} = \boldsymbol{diag}(25, 25)$, $\boldsymbol{\varepsilon} = 0.01$. Initial system configuration $q_0(0) = 0\text{rad}$, $q_1(0) = q_{\mathrm{m1}}(0) = 0.22\text{rad}$, $q_2(0) = q_{\mathrm{m2}}(0) = 0.79\text{rad}$. The expected trajectory of the system is as follows:

$$\begin{cases} q_{d0} = \pi/6\text{rad} \\ q_{d1} = \pi/8\text{- }\sin(\pi t/3)\text{rad} \\ q_{d2} = \cos(\pi t/3)\text{rad} \end{cases}$$

The continuous smooth static friction law proposed by Feeny was adopted to represent the friction characteristics of joints [21]:

$$\mathbf{f}(\dot{\mathbf{q}}) = \mathbf{f}_v\dot{\mathbf{q}} + (\mathbf{f}_c + \mathbf{f}_{cs}\text{sech}(\beta\dot{\mathbf{q}})\tanh(\alpha\dot{\mathbf{q}}))$$

The viscous friction factor $f_v = 0.00986$; $f_c = 0.743$ is coulomb friction factor; $f_{cs} = 3.99$ and $\alpha = 3.24$ structure Stribeck friction; $\beta = 0.799$ ensure the continuity of the equation when the relative sliding velocity is zero, $f(1) = 0$ (**Figure 9**).

The simulation results are as follows:

Figure 10 shows the tracking control of the carrier attitude Angle. It can be seen that without friction compensation control, the coupling effect of joint friction torque will cause interference to the carrier attitude, making it difficult to maintain the desired attitude Angle stably. Under the action of friction compensation control, the friction compensation term can effectively offset the coupling effect of joint friction

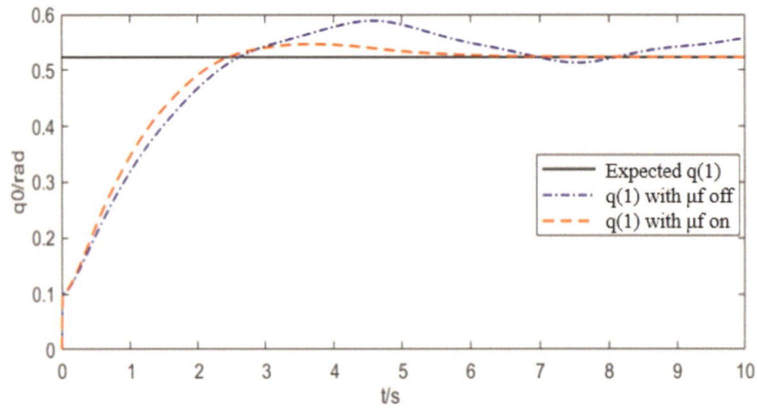

Figure 9.
Carrier attitude angle.

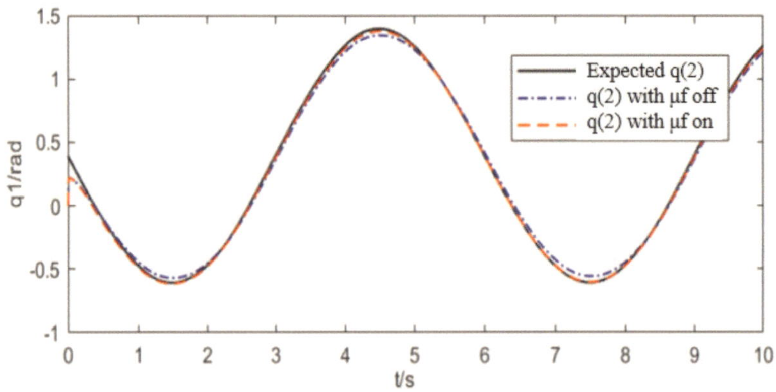

Figure 10.
Joint angle of arm 1.

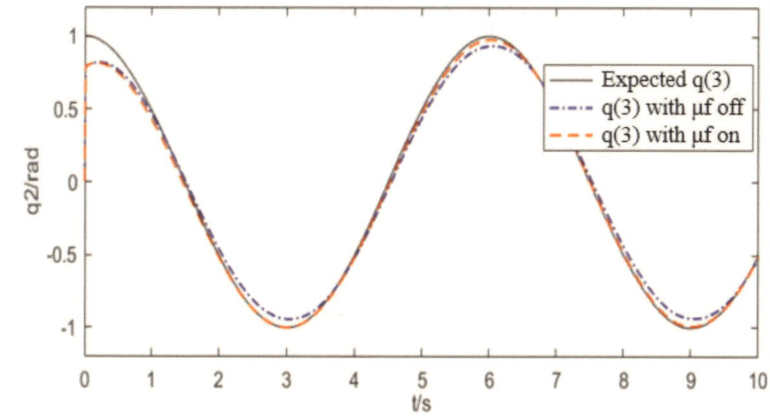

Figure 11.
Joint angle of arm 2.

torque on the carrier, and the carrier can maintain the desired attitude stably after 5 s. **Figure 10** and **Figure 11** show the joint angle tracking control situation. Due to the existence of friction factors, the joint angle tracking has obvious response hysteresis, the actual trajectory lags behind the expected trajectory, $|\dot{\mathbf{q}}| \le |\dot{\mathbf{q}}_d|$, and the real-time tracking error is large. After friction compensation control, the hysteresis phenomenon of joint angle tracking is effectively improved, and the real-time trajectory tracking performance is improved.

In **Figures 12** and **13**, τ_0, τ_1 and τ_2 represent the carrier attitude, joint 1 and joint 2 control input, respectively. By comparison, it can be seen that the controller designed in this chapter can effectively track the expected trajectory stably, the torque differential controller can effectively suppress the vibration caused by flexible joints, and the L2 gain robust controller based on friction compensation for the slow subsystem

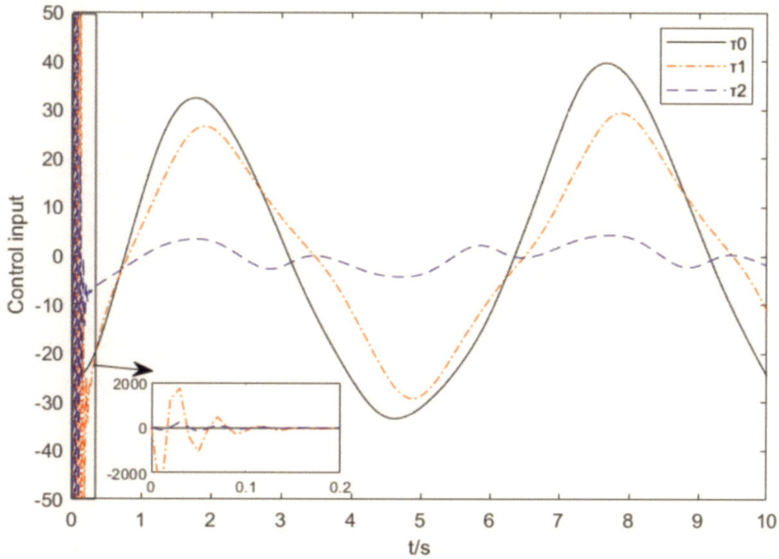

Figure 12.
Control input with friction compensation turn off.

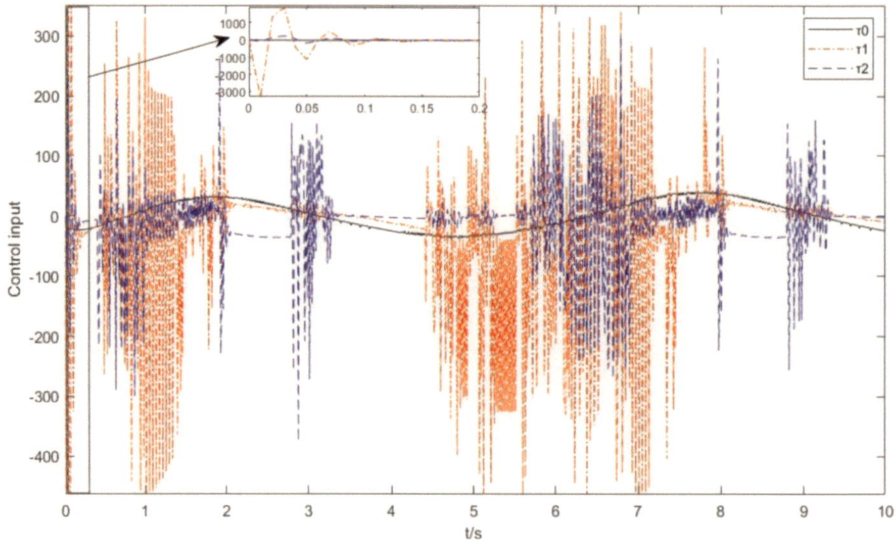

Figure 13.
Control input with friction compensation turn on.

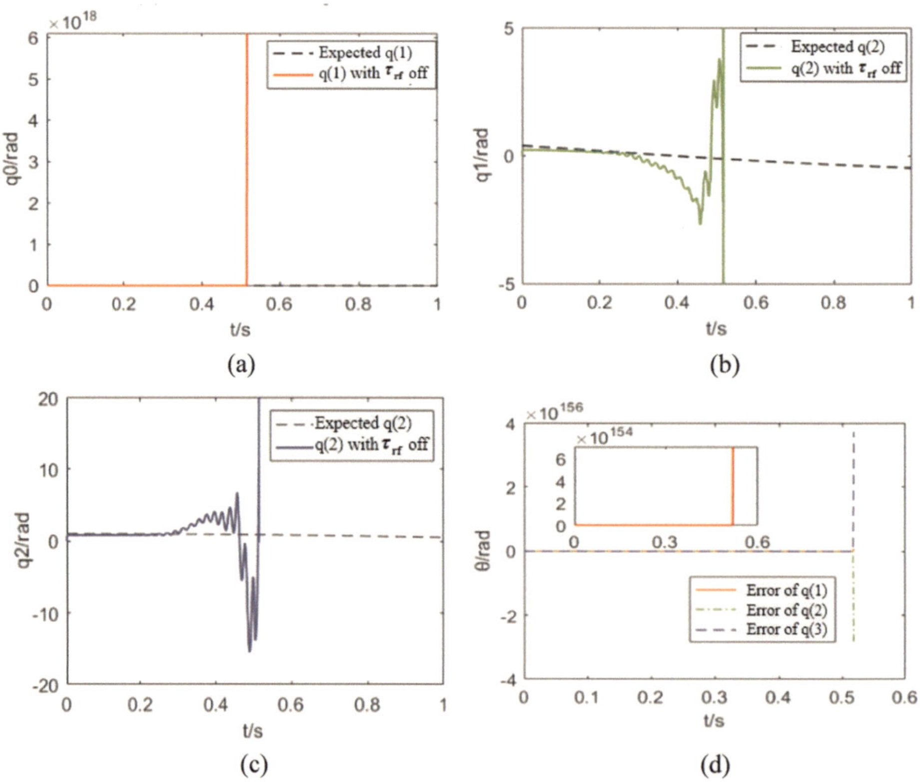

Figure 14.
Joint angle and error without vibration suppression. (a) Carrier attitude angle, (b) joint 1 angle, (c) joint 2 angle, (d) tracking error.

can effectively improve the response hysteria caused by joint friction torque and reduce the interference of friction torque coupling on the carrier attitude.

Figure 14 shows the joint Angle tracking curve and tracking error of the space robot system when the torque differential controller $\boldsymbol{\tau}_{rf}$ and friction compensation control $\mathbf{u}_f$ of the fast variable subsystem are turned off. By comparing **Figures 9-11** and **13**, it can be seen that after closing the torque differential controller $\boldsymbol{\tau}_{rf}$, the vibration caused by joint flexibility causes the instability of the carrier and joint motion, and the system oscillates and diverges, and the joint angle tracking control cannot be effectively implemented. It can be proved that the torque differential controller given by Eq. (58) can effectively suppress the vibration caused by joint flexibility.

4.1 Summary

In this section, the trajectory tracking control problem of flexible joint space robot system influenced by joint flexibility and joint friction torque is discussed. Based on joint compensation and singular perturbation method, the torque differential controller for the fast subsystem is designed. For the slow subsystem, the L2 gain controller based on the upper bound of friction is robust. The scheme can restrain the influence of friction factors well. Through the co-simulation, it can be seen that the space robot with floating base is different from the ground robot with fixed base. Its base attitude is affected by the coupling effect of joint friction torque. The controller designed can reduce the influence of joint coupling effect on the carrier attitude angle and effectively improve the response hysteretic problem caused by the friction factor.

The Lyapunov equation of the system is constructed, the convergence of the control law is proved by the stability theory, and the simulation proves that the controller designed in this chapter can effectively improve the response hysteretic problem caused by friction, and effectively reduce the influence of joint coupling factors on the attitude control of the carrier.

Author details

Xinye You
Wuhan Guangting Information Technology Co., Ltd, Wuhan, China

*Address all correspondence to: atlantyou@gmail.com

References

[1] Ding XL. Humanoid Two-Arm Robot Technology. Science Press; 2011

[2] Kane TR, Levinson DA. The use of Kane's dynamical equations in robotics. The International Journal of Robotics Research. 1983;**2**(3):1-21

[3] Albu-Schäffer A, Ott C, Hirzinger G. A Unified Passivity-based Control Framework for Position, Torque and Impedance Control of Flexible Joint Robots. Springer Berlin Heidelberg; 2007

[4] Vafa Z, Dubowsky S. The kinematics and dynamics of space manipulators: The virtual manipulator approach. International Journal of Robotics Research. 1990;**9**(4):3-21

[5] Roberson RE, Wittenburg J. A dynamical formalism for an arbitrary number of interconnected rigid bodies, with reference to the problem of satellite attitude control. Journal of Infectious Diseases. 1966;**207**(9):1359-1369

[6] Spong MW. Modeling and control of elastic joint robots. Asme Journal of Dynamic Systems, Measurement, and Control. 1987;**109**(4):310-319

[7] Huang XQ, Chen L. Finite time control and vibration suppression of a space robot with double flexible bars and dead zone. China Mechanical Engineering. 2019;**030**(010):1212-1218

[8] Chen JK, Ge LZ, Li RF. Robust controller design of robot flexible joint considering friction characteristics. Journal of Jilin University: Engineering Edition. 2014

[9] Meng G, Han LL, Zhang CF. Research progress and technical challenges of space robots journal of. Aviation. 2021; **42**(1):26

[10] Ye S, Butyrin S, Somova T, et al. Control of a free-flying robot at preparation for capturing a passive space vehicle. IFAC Papers On Line. 2018; **51**(30):72-76

[11] Ulrich S, Sasiadek JZ. Extended Kalman filtering for flexible joint space robot control. In: American Control Conference. IEEE; 2011

[12] Feeny B, Moon FC. Chaos in a forced dry-friction oscillator: Experiments and numerical modelling. Journal of Sound & Vibration. 1994;**170**(3):303-323

[13] Chen ZY, Chen L. Augmented adaptive control and joint vibration suppression of flexible joint space robot with flexible compensation. Acta Astronautica. 2013;**34**(12):1599-1604

[14] You XY, Chen L. Robust control of space robot based on disturbance observer under external disturbance journal of. Dynamics and Control. 2021

[15] Yoshisada M, Showzow TS, Kei S, et al. Trajectorycontrol of flexible manipulators on a free-flying space robot. IEEE Control Systems. 1992;**12**(3): 51-57

[16] Liu H, Huang Y, Shi W. Design of adaptive fuzzy controller for flexible link manipulator. IEEE International Conference on Industrial Technology, ICIT. 2008:1-4

[17] Wilson DG, Starr GP, Parker GG, et al. Robust control design for flexible-link/flexible-joint robots. In: IEEE International Conference on Robotics & Automation. IEEE; 2000

[18] Jing C, Li C. The fuzzy neural network control with H ∞ tracking

characteristic of dual-arm space robot after capturing a spacecraft. IEEE/CAA Journal of Automatica Sinica. 2018:99

[19] Shihabudheen KV, Jacob J. Composite control of flexible link flexible joint manipulator. In: India Conference (INDICON), 2012 Annual IEEE. IEEE; 2013

[20] Krenn R, Landzettel K, Kaiser C, et al. Simulation of the docking phase for the smart-OLEV satellite servicing mission. In: 9th International Symposium on Artificial Intelligence, Robotics and Automation in Space. DLR; 2008

[21] Chu M, Wu XY. Modeling and self–learning soft–grasp control for free-floating space manipulator during target capturing using variable stiffness method. IEE Access. 2018;**6**:7044-7054